Oxford International Primary Maths

Tony Cotton

Caroline Clissold

Linda Glithro

Cherri Moseley

Janet Rees

Language consultants:
John McMahon
Liz McMahon

4

OXFORD

UNIVERSITY PRESS

OXFORD

UNIVERSITY PRESS

Great Clarendon Street, Oxford, OX2 6DP, United Kingdom

Oxford University Press is a department of the University of Oxford. It furthers the University's objective of excellence in research, scholarship, and education by publishing worldwide. Oxford is a registered trade mark of Oxford University Press in the UK and in certain other countries

British Library Cataloguing in Publication Data
Data available

978-0-19-839462-4

10 9

Paper used in the production of this book is a natural, recyclable product made from wood grown in sustainable forests. The manufacturing process conforms to the environmental regulations of the country of origin.

Printed in Great Britain by Bell and Bain Ltd, Glasgow

FSC MIX Paper from responsible sources FSC® C007785

Acknowledgements

The publishers would like to thank the following for permissions to use their photographs:

Cover photo: Alex Staroseltsev/Shutterstock, P1a: Dominique Landau/Shutterstock, P1b: Typhoonski/Dreamstime.com, P1c: Daniel H. Bailey/Photolibrary/Getty Images, P1d: Atlantide Phototravel/Corbis/Image Library, P21a: Richmatts/iStock.com, P21b: Paul Harris/JAI/Corbis/Image Library, P55: Liane Cary/AGE Best/Visual Photos, P73: Chaowalit407/Dreamstime.com, P89a: Aaron Amat/Shutterstock, P89b: Roman Samokhin/Shutterstock, P89c: Svetlana Foote/Shutterstock, P89d: Eric Isselee/Shutterstock, P89e: Eric Isselee/Shutterstock, P89f: Eric Isselee/Shutterstock, P89g: Eric Isselee/Shutterstock, P89h: Eric Isselee/Shutterstock, P89i: Ultrashock/Shutterstock, P89j: Eric Isselee/Shutterstock, P89k: Christian Musat/Shutterstock, P89l: Eric Isselee/Shutterstock, P90a: Shutterstock, P90b: Eric Isselee/Shutterstock, P90c: Shutterstock, P90d: Shutterstock, P117a: Tulla/iStock.com, P117b: Ryan Mackay/Dreamstime.com, P117c: Sergiy Telesh/Shutterstock, P117d: Perutskyi Petro/Shutterstock, P137: Sergiy Palamarchuk/Shutterstock, P161: Vito Palmisano/Photographer's Choice/Getty Images, P177: M.C. Escher's "Reptiles" © 2013 The M.C. Escher Company-The Netherlands. All rights reserved. www.mcescher.com, P192a: MarchCattle/Shutterstock, P192b: Geargodz/iStock.com, P192c: Shutterstock, P207: Grey Square, 1923 (pen & ink and w/c), Kandinsky, Wassily (1866-1944)/Private Collection/Photo © Lefevre Fine Art Ltd., London/The Bridgeman Art Library.

Although we have made every effort to trace and contact all copyright holders before publication this has not been possible in all cases. If notified, the publisher will rectify any errors or omissions at the earliest opportunity.

Links to third party websites are provided by Oxford in good faith and for information only. Oxford disclaims any responsibility for the materials contained in any third party website referenced in this work.

The questions, example answers, marks awarded and/or comments that appear in this book were written by the author(s). In examination, the way marks would be awarded to answers like these may be different.

Contents

Unit **1** **Number and Place Value** I
Engage
1A Place value and partitioning ... 2
1B Counting on and back ... 7
1C Understanding and using decimal notation ... 10
1D Mental subtraction ... 14
1E Number sequences ... 17
Connect ... 21
Review ... 22

Unit **2** **Addition and Subtraction** 23
Engage
2A Addition to 100 and 1000 ... 24
2B Mental addition and subtraction ... 28
2C Mentally adding three or four small numbers ... 32
2D Mentally adding or subtracting 2-digit numbers ... 36
2E Adding or subtracting multiples of 10, 100, 1000 ... 40
2F Adding pairs of 3-digit numbers ... 43
2G Subtracting 2-digit and 3-digit numbers ... 47
Connect ... 51
Review ... 53

Unit **3** **Multiplication** 55
Engage
3A Multiplication tables and multiples ... 56
3B Multiplying 2-digit numbers ... 60
3C Changing the order of multiplying numbers ... 65
3D Multiplying and dividing 3-digit numbers by 10 ... 67
3E Doubling and halving ... 70
Connect ... 73
Review ... 74

Unit **4** **Division** 75
Engage
4A Dividing 2-digit numbers by a single-digit number ... 76
4B Rounding answers up or down ... 81
4C Multiplication and division as inverse operations ... 85
4D Ratio and proportion ... 89
Connect ... 93
Review ... 94

Unit **5** **Fractions** 95
Engage
5A Ordering and comparing fractions ... 96
5B Equivalent fractions ... 100
5C Using equivalence to order fractions ... 105
5D Mixed numbers ... 109
Connect ... 113
Review ... 115

Unit 6	**Decimals and Fractions**	**117**
	Engage	
	6A Decimals and tenths	118
	6B Using decimals for tenths and hundredths	122
	6C Equivalent fractions and decimals	126
	6D Finding fractions of shapes and numbers	130
	Connect	135
	Review	136

Unit 7	**Measurement, Area and Perimeter**	**137**
	Engage	
	7A Estimating, measuring and recording length	138
	7B Estimating, measuring and recording mass	142
	7C Estimating, measuring and recording capacity	146
	7D Using and reading scales	148
	7E Drawing rectangles and calculating perimeters	153
	7F Finding areas of rectangles	156
	Connect	159
	Review	160

Unit 8	**Time**	**161**
	Engage	
	8A Telling the time	162
	8B Timetables and calendars	167
	8C Measuring time intervals	172
	Connect	175
	Review	176

Unit 9	**Shape and Geometry**	**177**
	Engage	
	9A 2D shapes and classifying polygons	178
	9B 3D shapes	182
	9C Line symmetry	186
	9D 2D nets of 3D shapes	190
	Connect	192
	Review	194

Unit 10	**Position and Movement**	**195**
	Engage	
	10A Measuring angles	196
	10B Giving directions to follow a path	199
	10C Coordinates of a square on a grid	201
	Connect	207
	Review	208

Unit 11	**Handling Data**	**209**
	Engage	
	11A Collecting, presenting and interpreting data	210
	11B Comparing scales with different intervals	214
	11C Using Venn diagrams and Carroll diagrams	219
	Connect	223
	Review	225

Glossary		226

1 Number and Place Value

Engage

Olives

How do we count, write and order large numbers?

A thousand!

1A Place value and partitioning

The place or **position** of a **digit** in a number tells you its size or value.

Look at the number 2374:

The 2 has a value of 2000 –
there are 2 thousands.

Th	H	T	U
2	3	7	4

The 4 has a value of 4 units.

The 3 has a value of 300 –
there are 3 hundreds.

The 7 has a value of 70 – there
are 7 tens, which is seventy.

We read this number as two thousand three hundred **and** seventy-four.

When one of the places has no value we use a **zero as a placeholder**.

For example: In the number 3045 the zero shows that there are no hundreds.

We read this number as three thousand and forty-five.

I. Write these numbers in figures. The first one has been done for you.

a) Four thousand six hundred and thirty-four ___4634___

b) Six thousand one hundred and fifty-seven ___6157___

c) One thousand three hundred and twenty-two ___1322___

d) Five thousand four hundred and ninety-five ___5495___

e) Two thousand eight hundred and forty-nine ___2849___

f) Three thousand and sixty-nine ___3069___

g) Eight thousand three hundred and two ___8302___

h) Nine thousand and five ___9005___

2. Write these numbers in words. The first one has been done for you.

a) 7169 _Seven thousand one hundred and sixty-nine_

b) 4372 _four thousand three hundred and seventy-two_

c) 6723 _six thousand seven hundred and twenty-three_

d) 9821 _nine thousand eight hundred and twenty-one_

e) 3097 _three thousand ninety-seven_

f) 2409 _two thousand four hundred and nine_

g) 1560 _one thousad five hundred and sixty_

h) 5009 _five thousand and nine_

3. Look at these digits and answer the questions:

   7   4   9   1

a) What is the largest number that you can make with all four digits?

   _9741_

b) What is the largest even number that you can make with all four digits?

   _9714_

c) Using all four digits, make the smallest number possible: _1479_

d) Using all four digits, make the smallest even number possible: _1794_

- Check your answers with a partner.

4. When you know the value of the digits you can **partition** a number.

For example: 2135 = 2000 + 100 + 30 + 5

Complete these number statements:

a) 3621 = 3000 + _600_ + 20 + 1

b) 8516 = 8000 + 500 + 10 + 6

c) 4259 = 4000 + 200 + 50 + 9

d) 1857 = 1000 + 800 + 50 + 7

e) 6382 = 6000 + 300 + 80 + 2

f) 9174 = 9000 + 100 + 70 + 4

g) 7813 = 7000 + 800 + 10 + 3

5. Partition these numbers:

a) 1526 = $1000+500+20+6$

b) 4837 = $4000+800+30+7$

c) 3054 = $3000+50+4$ + 11

d) 7303 = $7000+300+0+3$

e) 6007 = $6000+7$

f) 8070 = $8000+70$

4

1A Place value and partitioning

I.

> 5632
>
> The underlined digit in this number is 5 thousands.

Write down the value of the digit that is <u>underlined</u>. Place the number on the number line.

a) 4<u>2</u>68

4 thousands

b) 3<u>2</u>79

2 hundreds

c) 6<u>7</u>05

7 hundreds

d) 254<u>1</u>

1 unit

e) 70<u>4</u>3

4 tens

2. Work with a partner. Look at these numbers:

40	300	70	2	100	8	90	7	800

Make four different 3-digit numbers. (For example, I can make 300 + 90 + 7 = 397.)

Place them on this number line.

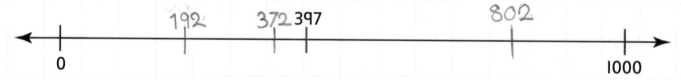

Round your numbers to the nearest 10.

___397___ is ___400___ rounded to the nearest 10.

___216___ is ___220___ rounded to the nearest 10.

___573___ is ___570___ rounded to the nearest 10.

___858___ is ___860___ rounded to the nearest 10.

___374___ is ___370___ rounded to the nearest 10.

3. Work with a partner. Look at these numbers:

2000	60	300	4	80	7	500	3	8000	50	5	5000

Make four different 4-digit numbers.
(For example, I can make 2000 + 500 + 60 + 7 = 2567.)

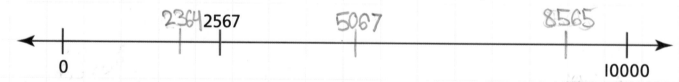

Round your numbers to the nearest 100.

___2567___ is ___2600___ rounded to the nearest 100.

___3457___ is ___3500___ rounded to the nearest 100.

___7831___ is ___7800___ rounded to the nearest 100.

___1183___ is ___1200___ rounded to the nearest 100.

___5629___ is ___5600___ rounded to the nearest 100.

1B Counting on and back

Discover

You can use **place value** to **count on** and **count back** in ones, tens, hundreds and thousands.

> For example: 5642 $\xrightarrow{+100}$ 5742 $\xrightarrow{+1000}$ 6742 $\xrightarrow{-1}$ 6741 $\xrightarrow{+10}$ 6751

1. Complete these steps:

   a) 2574 $\xrightarrow{+1000}$ 3574 $\xrightarrow{-10}$ 3564 $\xrightarrow{-100}$ 3464 $\xrightarrow{+1}$ 3465

   b) 4892 $\xrightarrow{+10}$ 4902 $\xrightarrow{+100}$ 5002 $\xrightarrow{-1000}$ 4002 $\xrightarrow{+100}$ 4102

   c) 7198 $\xrightarrow{+10}$ 7208 $\xrightarrow{+1000}$ 8208 $\xrightarrow{-100}$ 8108 $\xrightarrow{+1}$ 8109

   d) 6920 $\xrightarrow{+1000}$ 7920 $\xrightarrow{+100}$ 8020 $\xrightarrow{-10}$ 8010 $\xrightarrow{-1}$ 8009

   e) 1099 $\xrightarrow{+1}$ 1100 $\xrightarrow{+100}$ 1200 $\xrightarrow{+1000}$ 2200 $\xrightarrow{+10}$ 2210

2. Look at the numbers in the middle column of this table.

   - Count on and back to complete both sides of the grid.

     The first row shows an example.

−1000	−100	−10	−1	Number	+1	+10	+100	+1000
123	1123	1223	1233	1234	1235	1245	1345	2345
2150	3150	3250	3260	3261	3262	3272	3372	4372
2964	3964	4064	4074	4075	4076	4086	4186	5186
1078	2078	2178	2188	2189	2190	2200	2300	3300
6802	7802	7902	7912	7913	7914	7924	8024	9024
4988	5988	6088	6098	6099	6100	6110	6210	7210
6698	7698	7898	7908	7909	7910	7920	8020	9020

3. Use these number cards:

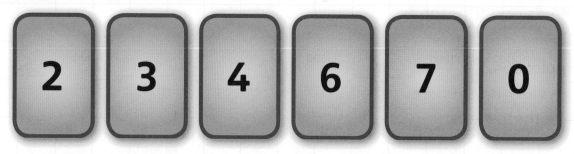

2 3 4 6 7 0

a) Make as many pairs of 4-digit numbers as you can that have
   a **difference** of 100.

2563
2463

6174
6074

7743
7643

3770
3670

4330
4230

b) Now make pairs of 4-digit numbers with a difference of 1000.

5670
4670

3376
2376

4764
3764

8234
7234

8236
7236

1B Counting on and back

Explore

1. A number adventure!

   Work with a partner.

   - Choose a 3-digit number.

   - Use a whiteboard each.

   - Take your number on this adventure:

Add 2000
Take away 2
Add 200
Take away 10

   $467 \rightarrow 2467 \rightarrow 2465 \rightarrow 2665 \rightarrow 2655$

   Check with your partner.

   Did you both reach the same final number?

   - Write your own number adventure using a 4-digit number.

   - Give your number adventure to your partner to test it.

   Try to make sure that you change each digit in your adventure.

2. Here are some computer games scores.

   Work out the difference between the start score and the new score.

Start score	New score	Difference
4560	4660	The new score is 100 more than the start score.
2913	3113	The new score is 200 more than the start score.
7521	9521	The new score is 2000 more than the start score.
1309	1349	The new score is 40 more than the start score.
3189	4289	The new score is 1100 more than the start score.
8732	8738	The new score is 6 more than the start score.
5689	6089	The new score is 400 more than the start score.

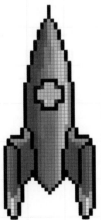

Discover

In a **decimal fraction** the **decimal point** separates the whole number from the **fraction**.

The first place after the point is for tenths.

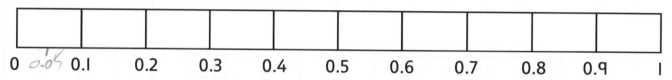

0   0.05 0.1   0.2   0.3   0.4   0.5   0.6   0.7   0.8   0.9   1

I. What part of each fraction is shaded?

  • Write your answers as a fraction and a decimal fraction.

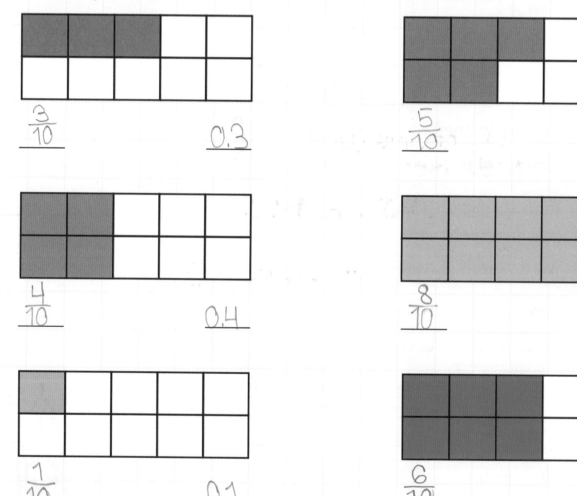

$\frac{3}{10}$ 0.3 $\frac{5}{10}$ 0.5

$\frac{4}{10}$ 0.4 $\frac{8}{10}$ 0.8

$\frac{1}{10}$ 0.1 $\frac{6}{10}$ 0.6

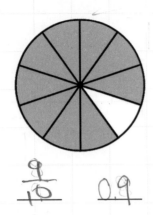

$\frac{9}{10}$ 0.9

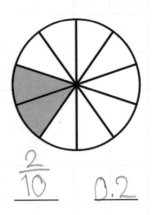

$\frac{2}{10}$ 0.2

2. Which decimal fraction is equal to a half? ___0.5___

3. Write each group of numbers in order, from smallest to largest.

a) $\frac{1}{2}$, 0.3, seven-tenths __0.3 $\frac{1}{2}$ seven-tenths__

b) 0.9, a half, three-tenths __three-tenths a half 0.9__

c) six-tenths, 1, 0.8 __six-tenths 0.8 1__

d) $\frac{4}{10}$, $\frac{1}{10}$, 0.5 __$\frac{1}{10}$ $\frac{4}{10}$ 0.5__

e) 2, 1.9, one and a half __a half one 1.9 2__

f) 3.8, 4.2, 4 __3.8 4 4.2__

g) four and three-tenths, 4.5, $3\frac{9}{10}$ __three-tenths $3\frac{9}{10}$ four 4.5__

4. Look at the number line.

   Write the number at each arrow in decimal form.

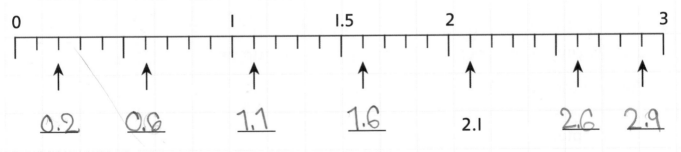

0.2   0.6   1.1   1.6   2.1   2.6   2.9

Explore

$$\frac{1.6\ 8}{\times 5\ 5} =$$

1. Look at the number line.

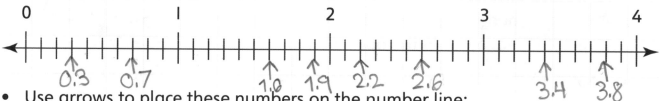

0   1   2   3   4

0.3   0.7   1.0 1.9 2.2 2.6   3.4 3.8

- Use arrows to place these numbers on the number line:

  a) 1.6   d) 2.2

  b) 3.4   e) 3.8

  c) 0.3   f) 1.9

- Add two more numbers of your own.

 2. Use a calculator.

> **Remember!**
>
> You can enter decimal fractions into a calculator.
>
> Key in '.1'. The calculator shows 0.1, a tenth.
>
> When you enter '+ 100 =', the calculator adds 100 to the number on the screen.
>
> Each time you press '=' the calculator repeats that **operation**, adding another 100.

- You can enter a start number into your calculator.

- You can then estimate how many times you need to press '=' after keying in the operator, to reach the final number.

- You can then use your calculator to check your estimate.

For example:

Start number	Final number	Operation	Estimate how many times you need to press '='	Actual number of presses required
10	70	+10	6	6

- Enter each start number from this table into your calculator.

- Look at the final number. How many times do you think you need to press '=' after keying in the operation to reach the final number?

- Record your estimate.

- Use the calculator to check your estimate.

- Record the actual number of presses required.

Start number	Final number	Operation	Estimate how many times you need to press '='	Actual number of presses required
0.1	0.4	+0.1	3	3
0.8	1.6	+0.1	9	8
4045	9045	+1000	5	5
582	632	+10	7	5
2541	3141	+100	10	6
178.7	179.7	+0.1	7	10
3102	8102	+1000	5	5
198.8	200.9	+0.1	12	29

1D Mental subtraction

Discover

1.

> Look at this subtraction:
>
> $479 - 198$
>
> **Step 1**
>
> I can change the sum to 479 – 200 and then adjust by adding 2.
>
> **Step 2**
>
> 479 – 200 is 279, then add 2. My answer is 281.

- Try these calculations using a similar method.

a) 632 – 297 = $297+3=300$ $632-300=332$ $332-3=329$

b) 854 – 199 = $199+1=200$ $854-200=654$ $654-1=653$

c) 617 – 398 = $398+2=400$ $617-400=217$ $217-2=215$

d) 501 – 202 = $202-2=200$ $501-200=301$ $301+2=303$

e) 962 – 403 = $403-3=400$ $962-400=562$ $562+3=565$

f) 720 – 302 = $302-2=300$ $720-300=420$ $420+2=422$

2. Explain to a partner how to do this calculation:

a) 469 + 203 = $203-3=200$ $469+200=669$ $669+3=672$

- Now try these:

b) 278 + 304 = $304 - 4 = 300$ $278 + 300 = 578$ $578 + 4 = 582$

c) 819 + 203 = $203 - 3 = 200$ $819 + 200 = 1019$ $1019 + 3 = 1022$

d) 568 + 404 = $404 - 4 = 400$ $568 + 400 = 968$ $968 + 4 = 972$

e) 109 + 602 = $602 - 2 = 600$ $109 + 600 = 709$ $709 + 2 = 711$

3.

Look at this addition:

$498 + 4 = 502$

Step 1
$498 + 2$ makes 500

$500 + 2$
My answer is 502.

- Do these calculations in the same way.

a) 397 + 5 = $397 + 3 = 400$ $400 + 2 = 402$

b) 796 + 7 = $796 + 4 = 800$ $800 + 3 = 803$

c) 2395 + 8 = $2395 + 5 = 2400$ $2400 + 3 = 2403$

d) 5799 + 6 = $5799 + 1 = 5800$ $5800 + 5 = 5805$

e) 8098 + 7 = $8098 + 2 = 8100$ $8100 + 5 = 8105$

$\begin{array}{r} 24 \\ 60 \\ \hline 00 \end{array}$

144×1440

Number and Place value

1D Mental subtraction

1. Calculate the missing numbers:

   a) 497 + ___7___ = 504

   b) 2199 + ___7___ = 2206

   c) 6397 + ___6___ = 6403

   d) 2098 + ___9___ = 2107

   e) 3399 + ___6___ = 3405

2. Here are some distances travelled by mini-buses:

498 km	7099 km	3499.5 km	169.2 km
398 km	3500 km	5799 km	3298.7 km

   a) Starting with the smallest, rewrite the distances in increasing order:

   169.2 km, 398 km, 498 km, 3298.7 km, 3499.5,
   3500 km, 5799 km, 7099 km

   The next day all the mini-buses make the same journey of 199 km.

   b) Write the total distance that each mini-bus has now travelled:

   368.2 km, 597 km, 697 km, 3497.7 km,
   3698.5 km, 3699 km, 5998 km, 7298 km

16

1E Number sequences

Discover

A **number sequence** is a sequence of numbers that follow a numerical **rule**.

Here are some examples:

2, 4, 6, 8, 10, 12, . . .   To find the next number, the rule is 'add 2'.

50, 45, 40, 35, . . .   To find the next number, the rule is 'subtract 5'.

1. You can write missing numbers in a number sequence.

> For example: 134, 135, <u>136</u>, 137, <u>138</u>, 139   (The rule is 'add 1'.)

- Look at these sequences.
- Write in the missing numbers.

  a) 702, 704, <u>706</u>, 708, <u>710</u>, 712

  b) 255, 260, 265, <u>270</u>, 275, <u>280</u>

  c) 345, 347, <u>349</u>, <u>351</u>, 353, <u>355</u>

  d) 123, 132, 141, <u>150</u>, <u>159</u>, <u>168</u>

  e) 3.2, 3.4, 3.6, <u>3.8</u>, <u>4.0</u>, 4.2, <u>4.4</u>

2. Here are the rules for some sequences.

The first **term** in each sequence is 1.

- Write the next four terms.

  For example: The rule is 'add 3'.   1, <u>4</u>, <u>7</u>, <u>10</u>, <u>13</u>

  a) The rule is 'add 100'.   1, <u>101</u>, <u>201</u>, <u>301</u>, <u>401</u>

  b) The rule is 'add 2000'.   1, <u>2001</u>, <u>4001</u>, <u>6001</u>, <u>8001</u>

  c) The rule is 'add 0.5'.   1, <u>1.5</u>, <u>2</u>, <u>2.5</u>, <u>3</u>

  d) The rule is 'subtract 0.1'.   1, <u>0.9</u>, <u>0.8</u>, <u>0.7</u>, <u>0.6</u>

3. • Write the next four numbers in each sequence.

• Explain the rule.

For example: 1450, 2450, 3450, 4450, 5450, 6450, 7450

The rule is add 1000.

a) 562, 572, 582, __592__ , __602__ , __612__ , __622__

The rule is __add 10__ .

b) 6409, 6309, 6209, __6109__ , __609__ , __599__ , __589__

The rule is __subtrat 100__ .

c) 6213, 5213, 4213, __3213__ , __2213__ , __1213__ , __213__

The rule is __subtract 1000__ .

d) 0.3, 0.6, 0.9, __1.2__ , __1.5__ , __1.8__ , __2.1__

The rule is __add 0.3__ .

e) 5.0, 4.8, 4.6, __4.4__ , __4.2__ , __4.0__ , __3.8__

The rule is __subtract 0.2__ .

In the sequences in questions 1–3 the rules used addition or subtraction.

You can use different **operations** for the rule of a sequence.

For example: **doubling** or **halving**, **multiplying** or **dividing**.

4. Look at these sequences.

• Write the next three numbers.

a) 1, 2, 4, 8, __16__ , __32__ , __64__

b) 100 000, 10 000, 1000, __100__ , __10__ , __1__ ,

c) 8000, 4000, 2000, __1000__ , __500__ , __250__

1E Number sequences

Explore

I. Write the first four terms for four sequences of your own.

Explain the rule for each sequence.

	Sequence	Rule
a)	132 , 232 , 332 , 432 , 532 , 632	The rule is: add 100
b)	5672 , 4672 , 3672 , 2672 , 1672 , 672	The rule is: subtract 1000
c)	100 , 200 , 400 , 800 , 1600 , 3200	The rule is: doubling
d)	12000 , 6000 , 3000 , 1500 , 750 , 375	The rule is: halfing

- Ask a friend to write the next terms of your sequences in the table.
- Check your partner's rules.

2. Here are some difficult sequences.

Work with a partner.

- Find the next three terms and the rule for these sequences.

> For example: 1, 2, 5, <u>14</u>, <u>41</u>, <u>122</u>
>
> The rule is: × 3 – 1

a) 1, 3, 7, 15, <u>31</u>, <u>63</u>, <u>127</u>

The rule is: ×2 +1

b) 1, 4, 10, 22, <u>46</u>, <u>94</u>, <u>190</u>

The rule is: ×2 +2

c) 1, 4, 9, 16, <u>25</u>, <u>36</u>, <u>49</u>, <u>64</u>, <u>81</u>

The rule is: square

d) 1, 1, 2, 3, 5, <u>8</u>, <u>13</u>, <u>21</u>, <u>34</u>, <u>55</u>

The rule is: FIBONACCI sequence

> The last two sequences have special names.

Square and Fibonacci sequence

1 Number and place value

Investigating the world's major rivers

The River Nile is the **longest** river in the world from its source to its delta on the Mediterranean Sea.

The River Amazon is the world's **biggest** river measured by the amount of water that flows down it. On average, about 20 swimming pools' worth of water flows out of the mouth of the Amazon every second.

The River Nile

Aerial view of the Amazon Delta

It's your turn!

- Research some rivers.

- Find ten of the world's longest rivers.
  They must be more than 1000 km long!

- For each river, find:

  a) the length in kilometres

  b) the countries or continent that the river flows through

  c) an interesting numerical fact about the river.

- Now order your rivers. Start with the longest.

- Present your findings in an interesting way.

- Talk to your teacher about your ideas.

1 Number and place value

- Write six different 4-digit numbers, between 3000 and 5000.

  Do not use more than one zero in each number.

  Choose three odd numbers and three even numbers.

- Write your numbers in order from smallest to largest:

Smallest ⟶ Largest

| 3200 | 3908 | 4372 | 4657 | 4878 | 4999 |

- Now mark your numbers on the blank number line as accurately as you can:

3000   5000

- Choose two of your numbers to start number sequences.

- For each number, write the first five terms and explain the rule for your sequences:

  1. __4999__, __4899__, __4799__, __4699__, __4599__

     The rule is __subtract 100__

  2. __3200__, __3250__, __3300__, __3350__, __3400__

     The rule is __add 50__

- Write a calculation using each of the other four numbers.

  Use what you have learned in this Unit. For example: adding or subtracting units, tens, hundreds or thousands, or partitioning one of your numbers.

  1. In this Unit I have learned Place value and partitioning.

  2. I learned counting on and back.

  3. I have learned how to use decimal notation.

  4. I also learned Mental Subtraction and Number Squences.

2 Addition and Subtraction

Engage

Every day we add up numbers.
When do we need to do this?

2A Addition to 100 and 1000

Discover

To reach 100 from a 2-digit number, you:

- use your knowledge of **number bonds** to 10
- find the next 10
- add the number of tens you need.

For example:

+3   +40

57 + 43 = 100

57   60   100

1. Use a number line to record the steps to reach 100.

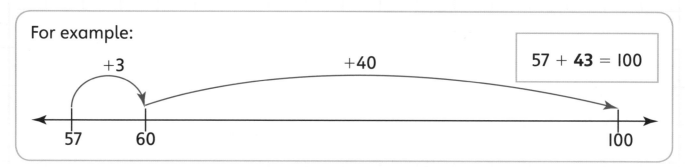

a) 24   +6   +70

24 +7 30   +30   100

b) 63

63 +3 70   +10   100

c) 87

87   90   + 80   100

d) 19   +7

19   20   +60   100

e) 32   +8

32 +9 40   +40   100

f) 51

51   60   100

2. How many different ways can you make 1000 using these numbers?

You can use a number more than once.

You can use two, three or four numbers in an **addition**.

| 50 | 100 | 150 | 200 | 250 | 300 | 350 | 400 | 450 | 500 |
| 550 | 600 | 650 | 700 | 750 | 800 | 850 | 900 | 950 |

For example: 100 + 900 = 1000

250 + 250 + 250 + 250 = 1000

50+950=1000   100+900=1000   150+850=1000   200+800=1000
250+750=1000   300+700=1000   350+650=1000   400+600=1000
450+550=1000   500+500=1000

3. Look at these ice creams and ice-lollies.

Can you see how much they cost?

You pay with a dollar bill. How much change do you get?

Cost	74c	42c	67c	18c	36c	59c	83c	21c
Change	26c	58c	33c	82c	64c	41c	17c	79c

2A Addition to 100 and 1000

1. These pairs of cards make 100.

   • Write the missing numbers on the blank cards.

a)	54	46	b)	73	27
c)	35	65	d)	28	72
e)	76	24	f)	69	31
g)	82	18	h)	47	53

2. Eight cars depart on a 1000 km journey.

This is how far they all travelled in one day:

How far does each car still have to go to reach 1000 km?

a) Car 1 travelled 250 km.   Distance still to go is ___750___ km

b) Car 2 travelled 450 km.   Distance still to go is ___550___ km

c) Car 3 travelled 300 km.   Distance still to go is ___700___ km

d) Car 4 travelled 650 km.   Distance still to go is ___350___ km

e) Car 5 travelled 550 km.   Distance still to go is ___450___ km

f) Car 6 travelled 350 km.   Distance still to go is ___650___ km

g) Car 7 travelled 700 km.   Distance still to go is ___300___ km

Ask a partner to check your answers.

3. Work with a partner.

You need a coloured pencil each and two dice.

- Take turns to roll the dice.

- Use the scores to make a 2-digit number.
  You can decide which order to use the digits.

- Work out a number pair that makes 100.

- Colour this number on the 100 square.
  If both possible numbers are already coloured, miss a go.

For example:
You throw a 2 and a 5.
You can choose 25 or 52.

For 25 you shade 75.

For 52 you shade 48.

1	2	3	4	5	6	7	8	9	10
11	12	13	14	15	16	17	18	19	20
21	22	23	24	25	26	27	28	29	30
31	32	33	34	35	36	37	38	39	40
41	42	43	44	45	46	47	48	49	50
51	52	53	54	55	56	57	58	59	60
61	62	63	64	65	66	67	68	69	70
71	72	73	74	75	76	77	78	79	80
81	82	83	84	85	86	87	88	89	90
91	92	93	94	95	96	97	98	99	100

Who has the most coloured squares at the end of the game? __orange__

Is it possible to roll every number? __NO__

- Explain your answer:

Because in a dice their is no 0
so we can't get 10,20,30,40 etc.

27

2B Mental addition and subtraction

Discover

1.

> To **add** 9, add 10 then **take away** 1.
>
> For example: 246 + 9 = 255   (Think 246 + 10 = 256, then 256 − 1 = 255)

- Complete this table:

		+9
a)	246	255
b)	572	581
c)	838	847
d)	153	162
e)	625	634
f)	497	506
g)	364	373
h)	789	798

2.

> To take away 11, take away 10 then take away 1
>
> For example: 627 − 11 = 616   (Think 627 − 10 = 617, then 617 − 1 = 616)

- Complete this table:

		−11
a)	627	616
b)	355	343
c)	784	773
d)	462	451
e)	279	268
f)	848	837
g)	191	180
h)	533	522

3. Work out these **additions**.

+	11	31	49	51	29
646	657	688	737	788	817
483	492	523	572	523	552
738	747	778	827	878	907

4. Use a number line to take away 99 from these numbers.

a) 165

b) 509

c) 237

d) 672

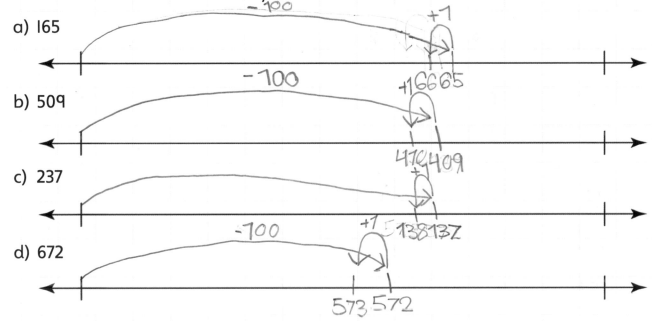

5. Use a number line to add 101 to these numbers.

a) 374

b) 824

c) 653

d) 405

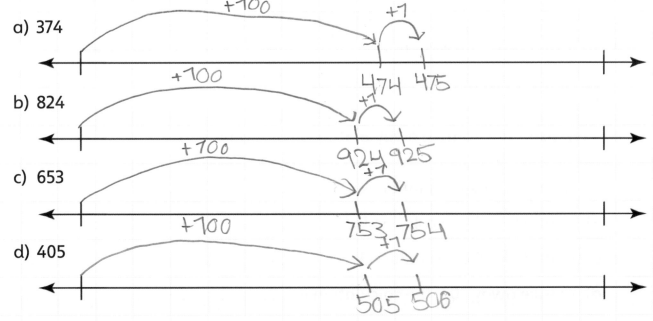

2B Mental addition and subtraction

1. Work with a partner.

   • Write six different 3-digit numbers.

   • One person adds 199 to each number.

   • The other person adds 201 to each number.

   • Record your answers in this table.

Our numbers	I added ___199___	My partner added ___201___
372	+ 571	573
536	+ 735	737
789	987	990
108	397	309
910	1109	1111
405	604	606

Compare your answers. What do you notice?

I add 200 to the number and take away 1.
My partner is add 200 to the numbers but
add 1.

2. What is an easy way to do these calculations? Talk to your partner.

   • Write the steps of each calculation:

   > For example: to take away 8, the calculation is − 10 + 2

   a) To add 13, the calculation is ___+10 +3___

   b) To add 99, the calculation is ___+100 −1___

   c) To add 197, the calculation is ___+200 −3___

   d) To take away 302, the calculation is ___−300 −2___

3. Work out the calculations and complete this table.

- One of you should use a calculator.

- The other should do the calculations mentally.

- Look at the + and − signs carefully.

	−201	−298	+402	+697
345	−200 −1	−300 +2	+400 +2	+700 −3
483	−200 −1	−300 +2	+400 +2	+700 −3
538	−200 −1	−300 +2	+400 +2	+700 −3

- Now compare your answers with your partner's.

 4. Work out the calculations and complete this table.

- One of you should do the calculations mentally.
  Swap roles from question 3.

- The other should use a calculator.

- Look at the + and − signs carefully.

	−199	+302	+598	+403
463	264	765	1061	866
257	58	559	855	660
721	522	1023	1319	1124

- Now compare your answers with your partner's.

5. Do you prefer using a calculator or a mental method? Why?

I prefer ___a mental method___

because ___a mental method chalnges my mind and a caloulatar doesn't.___

2C Mentally adding three or four small numbers

Discover

1.

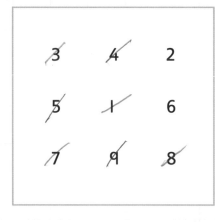

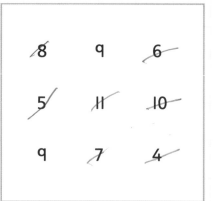

- Choose one number from each box to add together to make 20.

- Cross out the numbers as you use them.
  It is possible – but challenging – to use all the numbers!

- Write your calculations here:

5+10+5=20 9+3+8=20 7+7+6=20
7+9+10=20 3+6+11=20 4+9+7=20
8+8+4=20

2. The sum of the numbers in a magic square has the same total in every horizontal, vertical and diagonal line.

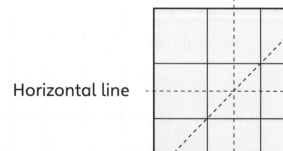

Put these numbers into the square so that the total in every horizontal, vertical and diagonal line is 150.

10	20	30
40	50	60
70	80	90

3. Work with a partner. Use a set of tens cards:

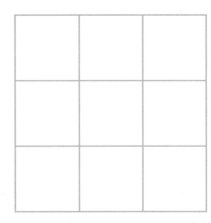

10 20 30 40 50 60 70 80 90

- One of you picks three cards.

- Both write the numbers in the table.

- Add the three numbers.

- Check that you and your partner have the same answer.

- Complete the table, taking turns to pick the numbers.

1st number	2nd number	3rd number	Total

Explore

1. How many different ways can you make the number 18 by adding three single-digit numbers?

2. Add these numbers mentally.

   Use pairs that make 10 or 20 to help you.

   > For example, adding 3, 8 and 17:
   >
   > $3 + 17 = 20$
   >
   > $20 + 8 = 28$
   >
   > so $3 + 8 + 17 = 28$

   a) 15, 2, 5, 9

   b) 8, 7, 4, 2

   c) 6, 4, 14

- Look at your number sentences.

- Underline pairs of numbers that add up to 10.

   d) 9, 9, 11

   > For example, using 1, 9 and 8:
   >
   > $\underline{1 + 9} + 8 = 18$

   e) 8, 9, 5, 3

   f) 9, 3, 7, 4

- Compare your answers with another student.

Did you both find all the possible answers?

g) 8, 7, 12, 3

3. Use three of these numbers.

3	6	9	15
16	7	13	2
11	4	18	5

Add them to make a total of less than 25.

For example: $7 + 13 + 2 = 22$

How many additions can you write in 5 minutes?

Try to be systematic.

Can you write 20 additions?

2D Mentally adding or subtracting 2-digit numbers

Discover

1. Add 56 to these numbers:

> You may want to **partition** 56.
>
> $56 = 50 + 6$
>
> Add the 50 and then add the 6 to reach the answer.

a) $45 + 56 =$

b) $67 + 56 =$

c) $82 + 56 =$

d) $39 + 56 =$

2. Add 68 to these numbers:

> You may want to partition 68.
>
> $68 = 60 + 8$
>
> Add the 60 and then add the 8 to reach the answer.

a) $37 + 68 =$

b) $76 + 68 =$

c) $29 + 68 =$

d) $95 + 68 =$

3. Use a number line method to complete these number sentences.

For example: To work out 74 − 28, **count on** from 28 to 74.

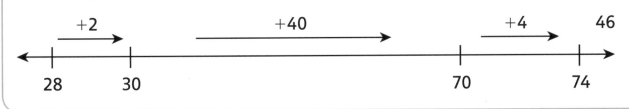

a) 65 − 27

b) 80 − 43

c) 74 − 35

d) 82 − 46

4. Each brick is the sum of the two bricks that it stands on.

- Write the missing numbers:

a)

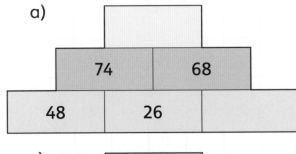

b)

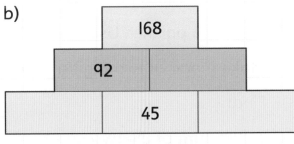

c)

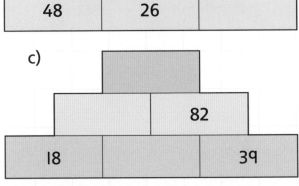

d)

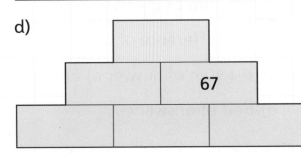

2D Mentally adding or subtracting 2-digit numbers

Explore

1. • Choose a number from each circle.

   • Use your numbers to write and solve an **addition** number sentence.

   • Use your numbers to write and solve a **subtraction** number sentence.

   • Choose two different numbers.

   • Repeat the above.

   • Continue until you have used all the numbers.

2. Work with a partner. Use number cards.

   • Make a pair of two-digit numbers.

   Find:

   a) the **sum** of the two numbers

   b) the **difference** of the two numbers

   c) the sum of answer a) and answer b)

   d) half of answer c).

- First complete this example for number cards 4, 5, 7 and 6.

  Pair of 2-digit numbers: 45 and 76

  a) 45 + 76 = _____

  b) 76 − 45 = _____

  c) _____

  d) _____

- Now you try:

- Compare your answers with your partner.

  What do you notice?

  _____

  _____

- Try more pairs of numbers:

Is the result the same every time? _____

2E Adding or subtracting multiples of 10, 100, 1000

Discover

I. Complete these sets of calculations.

Say the numbers quietly to yourself as you complete the list.

a)	$16 - 5 = \underline{\hspace{2cm}}$	$160 - 50 = \underline{\hspace{2cm}}$	$1600 - 500 = \underline{\hspace{2cm}}$
b)	$8 + 23 = \underline{\hspace{2cm}}$	$80 + 230 = \underline{\hspace{2cm}}$	$800 + 2300 = \underline{\hspace{2cm}}$
c)	$34 - 6 = \underline{\hspace{2cm}}$	$\underline{\hspace{1.5cm}} - \underline{\hspace{1.5cm}} = \underline{\hspace{1.5cm}}$	$\underline{\hspace{1.5cm}} - \underline{\hspace{1.5cm}} = \underline{\hspace{1.5cm}}$
d)	$7 + 19 = \underline{\hspace{2cm}}$	$\underline{\hspace{1.5cm}} + \underline{\hspace{1.5cm}} = \underline{\hspace{1.5cm}}$	$\underline{\hspace{1.5cm}} + \underline{\hspace{1.5cm}} = \underline{\hspace{1.5cm}}$
e)	$4 + 47 = \underline{\hspace{2cm}}$	$\underline{\hspace{1.5cm}} + \underline{\hspace{1.5cm}} = \underline{\hspace{1.5cm}}$	$\underline{\hspace{1.5cm}} + \underline{\hspace{1.5cm}} = \underline{\hspace{1.5cm}}$
f)	$53 - 9 = \underline{\hspace{2cm}}$	$\underline{\hspace{1.5cm}} - \underline{\hspace{1.5cm}} = \underline{\hspace{1.5cm}}$	$\underline{\hspace{1.5cm}} - \underline{\hspace{1.5cm}} = \underline{\hspace{1.5cm}}$
g)	$3 + 68 = \underline{\hspace{2cm}}$	$\underline{\hspace{1.5cm}} + \underline{\hspace{1.5cm}} = \underline{\hspace{1.5cm}}$	$\underline{\hspace{1.5cm}} + \underline{\hspace{1.5cm}} = \underline{\hspace{1.5cm}}$
h)	$85 + 7 = \underline{\hspace{2cm}}$	$\underline{\hspace{1.5cm}} + \underline{\hspace{1.5cm}} = \underline{\hspace{1.5cm}}$	$\underline{\hspace{1.5cm}} + \underline{\hspace{1.5cm}} = \underline{\hspace{1.5cm}}$

2. Solve these calculations, starting each one with the number in the blue shape.

For example: 5600 + 170 = 5770

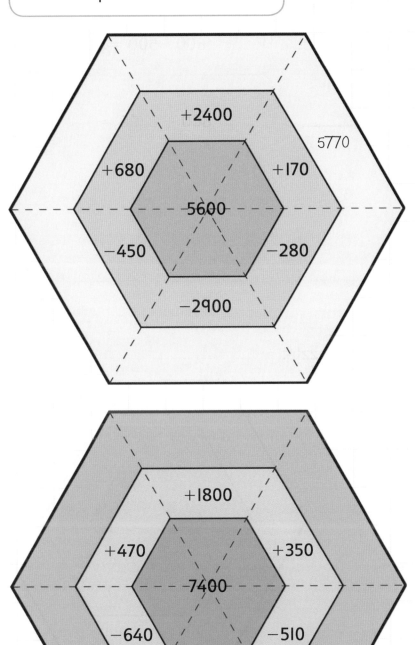

+2400

5770

+680

+170

5600

−450

−280

−2900

+1800

+470

+350

7400

−640

−510

−3800

Explore

1. The total of two numbers is less than 300.

   The difference is 60.

   The answer is a multiple of 10.

   How many number sentences can you write that fit these three facts?

   > For example: $140 + 80 = 220$
   >
   > $140 - 80 = 60$

2. Design your own hexagon number puzzle:

   - Write a 4-digit multiple of 100 in the blue hexagon.

   - Write additions or subtractions of multiples of 10 and 100 in the middle hexagon.

   - Work out the calculations.

   - Write your answers in the outer hexagon.

   - Check your solutions with another student.

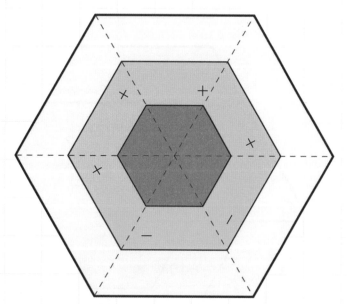

2F Adding pairs of 3-digit numbers

Discover

Remember: Think about the numbers and what you are doing!
Estimate your answer first.

I. Look at this table with the costs of different coloured bikes.

| $179 | $219 | $158 | $259 | $182 | $237 |

A red bike costs $179.

A blue bike costs $182.

A red bike and a blue bike cost

$$
\begin{array}{r}
179 \\
+ \quad 182 \\
\hline
\end{array}
$$

So, a red bike and a blue bike cost $ _____

Now find the cost of:

a) a green bike and a yellow bike

b) a black bike and a red bike

c) a white bike and a blue bike

g) the two most expensive bikes.

d) two red bikes

2. Add these pairs of 3-digit numbers.

Decide the method to use for each calculation.

You can do two of the calculations mentally.

a) 432 + 541

e) a white bike and a green bike

b) 278 + 175

c) 349 + 264

d) 337 + 198

e) 526 + 364

f) the two cheapest bikes

f) 189 + 427

g) 302 + 519

h) 154 + 457

2F Adding pairs of 3-digit numbers

1. Work with a partner. Use these numbers:

   Show your working. Find two numbers with:

428	236
127	509
105	397
261	129
366	487
318	273

   a) the largest even **total**

   _____

   _____

   b) the smallest odd total

   _____

   _____

   c) the total closest to 500

   _____

   _____

* Make up and solve two challenges of your own.

* Find the solutions.

* Ask another pair to solve your challenges!

   d) _____

   _____

   _____

   e) _____

   _____

   _____

2. Work with a partner.

- Roll three dice to give three different digits.

- Use the numbers to make all the possible 3-digit numbers.

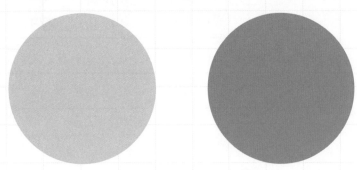

- Write your 3-digit numbers in the blue circle.

- Roll the dice again.

- Use the new numbers to make all the possible 3-digit numbers.

- Write these 3-digit numbers in the red circle.

- Use one number from the red circle and one number from the blue circle to work out:

  a) the largest even **total**   b) the smallest odd total

  c) the total closest to 500.

Make up and solve two challenges of your own.

d)   e)

Ask another pair to solve your challenges!

Discover

1.

> I know 147 + 138 = 285
>
> so I also know that:
>
> 138 + 147 = 285   285 − 147 = 138   285 − 138 = 147

- Write three more number sentences for each of these:

   a) 243 + 649 = 892

   b) 700 − 278 = 422

   c) 329 + 495 = 824

2. Eight friends have 850c each.

   Each friend buys a different chocolate bar at the price shown.

   How much money does each person have left?

   Use **counting on** or a number line method.

   a) 135c

   b) 271c

c) 359c

d) 401c

e) 524c

f) 646c

g) 518c

h) 239c

3. Here are the heights of seven hills in metres:

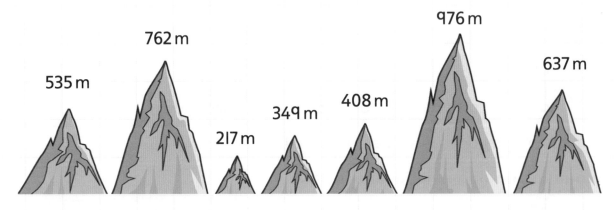

535 m
762 m
217 m
349 m
408 m
976 m
637 m

a) Work out the difference in height between two of the hills.

b) Repeat for a different pair of hills.

2G Subtracting 2-digit and 3-digit numbers

Explore

I.

	1st plane	2nd plane	3rd plane	4th plane
Capacity	269 seats	172 seats	114 seats	345 seats

Here are the numbers of seats sold for the Saturday and Sunday flights:

	Plane 1		Plane 2		Plane 3		Plane 4	
	Seats sold	Seats left	Seats sold	Seats left	Seats sold	Seats left	Seats sold	Seats left
Saturday flight	94		78		75		129	
Sunday flight	89		97		87		158	

How many seats are left for the Saturday and Sunday flights?

- Choose a method to use.

- Complete the table.

- Show your working:

2. Solve these three subtraction calculations.

Choose which method to use.

a) $352 - 176 =$

b) $\$241 - \$156 =$

c) $720\,kg - 352\,kg =$

3. Make up a story to fit the numbers for each number sentence in question 2.

a) _____

_____

_____

b) _____

_____

_____

c) _____

_____

_____

2 Addition and subtraction

Connect

English journeys

This map shows the positions
of five cities in England:

This chart gives information about the direct distances and
travel times between these cities:

	Distance (kilometres)	Hours	Minutes
Norwich to Oxford	272	3	35
Norwich to Leicester	190	2	40
Norwich to London	190	2	40
Norwich to Cambridge	105	1	25
London to Leicester	167	2	20
London to Oxford	100	1	25
London to Cambridge	100	1	25
Leicester to Oxford	122	1	40
Leicester to Cambridge	117	1	35
Cambridge to Oxford	166	2	15

Imagine you want to travel to more than two cities.

> For example: You want to go from Norwich to Cambridge and then to Oxford.
>
> Or you want to go from London to Oxford and then to Leicester.

- Write some journeys like these involving three or four cities:

  1. _____

  2. _____

  3. _____

  4. _____

- Use your journeys to answer these questions:

  a) What is the total distance of your journey in kilometres?

  b) How many minutes long is your journey?

  c) How much longer (in time) is your journey than a direct journey
     from the start city to the end city?

  _____

  _____

  _____

  _____

A final challenge! (Ask your teacher if you can use a calculator.).

Can you find the shortest route (in distance) to visit all five cities?

_____

_____

Can you find the longest route (in distance) to visit all five cities?

52

_____

_____

2 Addition and subtraction

Review

Here are some details for a film shown at the local cinema:

	Number of tickets sold	Ticket sales ($)	Sales of snacks ($)
Monday	323	3230	1432
Tuesday	415	4150	2487
Wednesday	489	4890	2356
Thursday	623	6230	3709

a) How many tickets were sold altogether on Monday and Tuesday? _____

_____

_____

b) Which day were the most tickets sold? _____

c) How much does a ticket cost? _____

_____

d) How much more money was spent on snacks on Wednesday than on Monday?

_____

- Use the information in the table.

- Make up five more addition and subtraction questions.

e) _____

f) _____

g) _____

h) _____

i) _____

3 Multiplication

I see lots of images.

The picture repeats itself.

The same picture over and over again.

3A Multiplication tables and multiples

Discover

1.

1	2	3	4	5	6	7	8	9	10
11	12	13	14	15	16	17	18	19	20
21	22	23	24	25	26	27	28	29	30
31	32	33	34	35	36	37	38	39	40
41	42	43	44	45	46	47	48	49	50
51	52	53	54	55	56	57	58	59	60
61	62	63	64	65	66	67	68	69	70
71	72	73	74	75	76	77	78	79	80
81	82	83	84	85	86	87	88	89	90
91	92	93	94	95	96	97	98	99	100

- Colour the **multiples** of 2 in yellow.

- Write a description of the pattern.

  Here are some useful words: column, vertical, alternate, even.

  _____

  _____

- On the same square, colour the multiples of 4 in orange.

  Can you see a new pattern?

- Describe the new pattern.

  _____

  _____

- Now count and colour the multiples of 8 in red.

- Describe this pattern.

  _____

  _____

2. The multiples of 3, 6 and 9 give a new family of patterns.

1	2	3	4	5	6	7	8	9	10
11	12	13	14	15	16	17	18	19	20
21	22	23	24	25	26	27	28	29	30
31	32	33	34	35	36	37	38	39	40
41	42	43	44	45	46	47	48	49	50
51	52	53	54	55	56	57	58	59	60
61	62	63	64	65	66	67	68	69	70
71	72	73	74	75	76	77	78	79	80
81	82	83	84	85	86	87	88	89	90
91	92	93	94	95	96	97	98	99	100

- Colour multiples of 3 in green, multiples of 6 in blue and multiples of 9 in purple.

- Describe the patterns you can see for multiples of 3, 6 and 9.

  Useful words are: diagonal, alternate, left, right, sloping.

3 _____

_____

_____

6 _____

_____

_____

9 _____

_____

_____

3A Multiplication tables and multiples

Explore

1. Complete these multiplications:

   a) $6 \times 8 = $ _____

   b) $5 \times 9 = $ _____

   c) $8 \times 0 = $ _____

   d) $9 \times 3 = $ _____

   e) $4 \times 7 = $ _____

   f) $10 \times 5 = $ _____

   g) $2 \times 9 = $ _____

   h) $8 \times 8 = $ _____

2. Write the missing numbers:

   a) $5 \times$ _____ $= 20$

   b) $6 \times$ _____ $= 36$

   c) _____ $\times 4 = 32$

   d) $8 \times$ _____ $= 24$

   e) $3 \times$ _____ $= 21$

   f) _____ $\times 9 = 54$

   g) _____ $\times 10 = 90$

   h) $2 \times$ _____ $= 14$

3. Find the **smallest** number that is:

   a) a multiple of 2 and 3 _____

   b) a multiple of 3 and 5 _____

   c) a multiple of 4 and 5 _____

   d) a multiple of 10 and 4 _____

   e) a multiple of 9 and 5 _____

   f) a multiple of 4 and 9 _____

   g) a multiple of 3 and 10 _____

   h) a multiple of 6 and 5 _____

4. Read these statements.

   Are they true or false?

   • Circle the correct answer.

   All multiples of 10 are also multiples of 5.   True/False

   All multiples of 4 are also multiples of 3.   True/False

   All multiples of 6 are also multiples of 3.   True/False

   All multiples of 4 are even numbers.   True/False

   All multiples of 3 are odd numbers.   True/False

   All multiples of 10 have 0 as the units digit.   True/False

   24 is a multiple of 3, 4 and 6.   True/False

   30 is a multiple of 3, 5 and 6.   True/False

5. Use the digits in this box, and as many 0s as you like to make multiples of:

| 3 | 5 | 7 | 2 | 6 | 4 |

a) 5 _____

b) 10 _____

c) 100 _____

You should find at least five multiples in each line. Include 3-digit numbers in your answers.

6. Add three numbers to each section of these Venn diagrams:

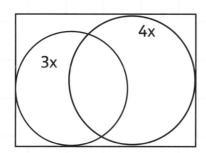

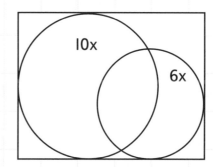

7. The multiples of 4 are   4, 8, 12, 16, 20, 24, 28, 32, 36, 40, ...

The digits in the units position are: 4, 8, 2, 6, 0, 4, 8, 2, 6, 0, ...

Can you see a pattern?

_____

_____

Do you think this pattern will continue?

• Test the next few numbers.

_____

_____

What happens when you try other multiple tables?

3B Multiplying 2-digit numbers

Discover

This calculation uses **partitioning** and **recombining**:

$32 \times 4 = (30 \times 4) + (2 \times 4) = 120 + 8 = 128$

1. **Partition** the 2-digit number to calculate each answer:

   a) $56 \times 3 = (\rule{1cm}{0.4pt} \times \rule{1cm}{0.4pt}) + (\rule{1cm}{0.4pt} \times \rule{1cm}{0.4pt}) = \rule{1cm}{0.4pt} + \rule{1cm}{0.4pt} = \rule{1cm}{0.4pt}$

   b) $44 \times 5 = (\rule{1cm}{0.4pt} \times \rule{1cm}{0.4pt}) + (\rule{1cm}{0.4pt} \times \rule{1cm}{0.4pt}) = \rule{1cm}{0.4pt} + \rule{1cm}{0.4pt} = \rule{1cm}{0.4pt}$

   c) $28 \times 6 = (\rule{1cm}{0.4pt} \times \rule{1cm}{0.4pt}) + (\rule{1cm}{0.4pt} \times \rule{1cm}{0.4pt}) = \rule{1cm}{0.4pt} + \rule{1cm}{0.4pt} = \rule{1cm}{0.4pt}$

   d) $53 \times 9 = (\rule{1cm}{0.4pt} \times \rule{1cm}{0.4pt}) + (\rule{1cm}{0.4pt} \times \rule{1cm}{0.4pt}) = \rule{1cm}{0.4pt} + \rule{1cm}{0.4pt} = \rule{1cm}{0.4pt}$

   e) $87 \times 2 = (\rule{1cm}{0.4pt} \times \rule{1cm}{0.4pt}) + (\rule{1cm}{0.4pt} \times \rule{1cm}{0.4pt}) = \rule{1cm}{0.4pt} + \rule{1cm}{0.4pt} = \rule{1cm}{0.4pt}$

   f) $34 \times 6 = (\rule{1cm}{0.4pt} \times \rule{1cm}{0.4pt}) + (\rule{1cm}{0.4pt} \times \rule{1cm}{0.4pt}) = \rule{1cm}{0.4pt} + \rule{1cm}{0.4pt} = \rule{1cm}{0.4pt}$

   g) $92 \times 3 = (\rule{1cm}{0.4pt} \times \rule{1cm}{0.4pt}) + (\rule{1cm}{0.4pt} \times \rule{1cm}{0.4pt}) = \rule{1cm}{0.4pt} + \rule{1cm}{0.4pt} = \rule{1cm}{0.4pt}$

2. This calculation uses the **grid method** to show the answer to 6×47:

×	40	7
6	240	42

   $240 + 42 = 282$

   Answer: $6 \times 47 = 282$

   Use the grid method to work out these multiplications.

   a) 4×58

   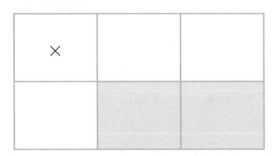

   $\rule{2cm}{0.4pt} + \rule{2cm}{0.4pt} = \rule{2cm}{0.4pt}$

   Answer: $4 \times 58 = \rule{3cm}{0.4pt}$

b) 3 × 85

_____ + _____ = _____

Answer: 3 × 85 = _____

d) 9 × 35

_____ + _____ = _____

Answer: 9 × 35 = _____

c) 5 × 64

_____ + _____ = _____

Answer: 5 × 64 = _____

e) 6 × 73

_____ + _____ = _____

Answer: 6 × 73 = _____

3.

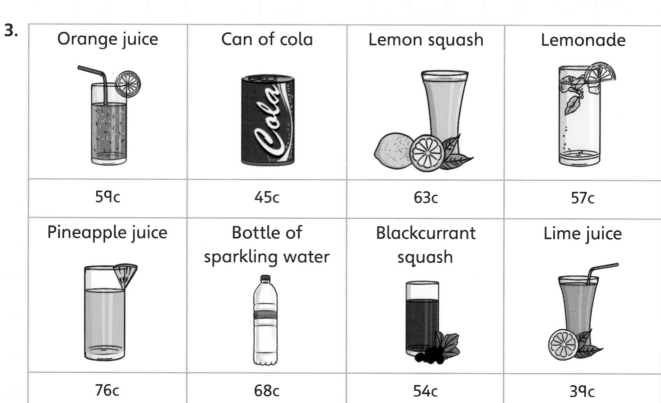

Orange juice	Can of cola	Lemon squash	Lemonade
59c	45c	63c	57c
Pineapple juice	Bottle of sparkling water	Blackcurrant squash	Lime juice
76c	68c	54c	39c

- Calculate the cost of these orders in the café.
- Use a method of your choice.

   a) 6 glasses of orange juice

   b) 5 cans of coke

   c) 7 glasses of lemon squash

   d) 4 glasses of lemonade

   e) 3 glasses of pineapple juice

   f) 8 bottles of sparkling water

   g) 9 glasses of blackcurrant squash

   h) 2 glasses of lime juice

3B Multiplying 2-digit numbers

I.

Triangle

Square

Pentagon

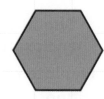
Hexagon

Heptagon

Octagon

Nonagon

Decagon

A zoo wants to build new enclosures of different shapes for the animals.

How many sides does each enclosure have?

- Write your answer in the third column of the table.

How much fencing do they need to buy?

- Write your answer in the fourth column of the table.

The fencing costs $10 a metre.

How much does fencing cost for each enclosure?

- Write your answer in the final column on the table.

Shape of enclosure	Length of side	Number of sides	Length of fencing required	Cost of fencing
Equilateral triangle	87 m			
Square	67 m			
Pentagon	56 m			
Hexagon	49 m			
Heptagon	43 m			
Octagon	36 m			
Nonagon	27 m			
Decagon	19 m			

2. • Use the digits in this box:

   5   7   4   3   8   2

   • Make up ten multiplication sums of the type

   TU × U =

   For example: 57 × 2 = 114

Discover

1.

This **array** shows 24 as 3 × 8:

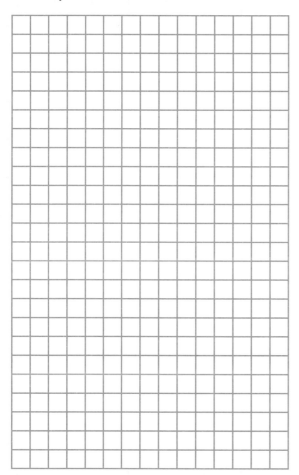

- Draw as many different arrays as you can for 24:

2. Make up four different multiplications with the answer:

a) 36 _____

_____

b) 60 _____

_____

c) 100 _____

_____

3. Draw arrays for 11 and 17.

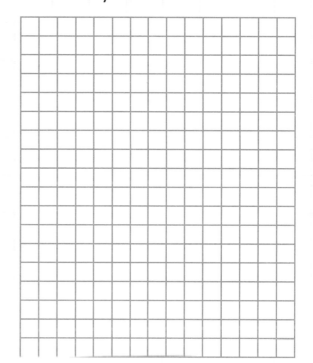

4. Change these calculations to solve them by **doubling** (×2):

For example:

16 × 4 becomes 32 x 2 = 64

a) 27 × 4 becomes _____ × 2 = _____

b) 13 × 6 becomes _____ × 2 = _____

c) 62 × 4 becomes _____ × 2 = _____

d) 24 × 6 becomes _____ × 2 = _____

Multiplication

65

3C Changing the order of multiplying numbers

1. Complete these multiplications in two ways.

   • Underline the numbers that you multiplied first.

   a) 5 × 3 × 2 = _____ × _____ = _____

      5 × 3 × 2 = _____ × _____ = _____

   b) 10 × 4 × 3 = _____ × _____ = _____

      10 × 4 × 3 = _____ × _____ = _____

   c) 9 × 2 × 5 = _____ × _____ = _____

      9 × 2 × 5 = _____ × _____ = _____

   d) 2 × 7 × 5 = _____ × _____ = _____

      2 × 7 × 5 = _____ × _____ = _____

   e) 5 × 10 × 4 = _____ × _____ = _____

      5 × 10 × 4 = _____ × _____ = _____

   f) 4 × 6 × 3 = _____ × _____ = _____

      4 × 6 × 3 = _____ × _____ = _____

   Did you find one way easier?

   For each one, tick ✓ the way you found easier or quicker.

2. Find one or more ways to write these multiplications and solve them.

   a) 8 × 5 × 4 = _____   c) 4 × 6 × 3 = _____

   _____   _____

   _____   _____

   b) 9 × 6 × 5 = _____   d) 2 × 8 × 5 × 3 = _____

   _____   _____

   _____   _____

Discover

I. Work with a partner.

Take turns to use the calculator.

When you use the calculator, key in the whole calculation.
For example: '341 × 10 ='.

Then copy the answer from the screen.

When you do the sum mentally, write the answer as soon as
you have worked it out.

× 10	I did these using a calculator (my partner did them mentally)	I did these mentally (my partner used the calculator)
341		
532		
612		
764		
492		
351		
639		
278		

Which method do you find quicker? _____

Multiplication

67

2. What do you think the answer to the multiplication 23 × 10 is? _____

What do you think that answer is multiplied by 10? _____

- Try these multiplications on the calculator. Were you correct?

  23 × 10 = _____

  _____ × 10 = _____

- Try the same calculations with 2-digit and 3-digit numbers of your choice.

  Write your answers:

3D Multiplying and dividing 3-digit numbers by 10

When you **multiply** by ten each digit becomes 10 **times** bigger.

The units become tens.

The tens become hundreds.

The hundreds become thousands.

For example: $456 \times 10 = 4560$

1. Complete these:

    a) $234 \times 10 =$ _____

    b) $171 \times 10 =$ _____

    c) $507 \times 10 =$ _____

    d) $962 \times 10 =$ _____

    e) $743 \times 10 =$ _____

    f) $608 \times 10 =$ _____

When you **divide** by ten each digit becomes 10 times smaller.

The thousands become hundreds.

The hundreds become tens.

The tens become units.

For example: $1570 \div 10 = 157$

2. Complete these:

    a) $340 \div 10 =$ _____

    b) $540 \div 10 =$ _____

    c) $650 \div 10 =$ _____

    d) $410 \div 10 =$ _____

    e) $250 \div 10 =$ _____

    f) $730 \div 10 =$ _____

    g) $890 \div 10 =$ _____

3. How many cents are there in these amounts of dollars?

    a) $17 _____

    b) $42 _____

    c) $97 _____

    d) $61 _____

    e) $39 _____

Multiplication

69

3E Doubling and halving

Discover

Remember: doubling is the same as multiplying by 2.

1. **Double** these numbers by doubling the tens, doubling the units, then combining.

> For example:
>
> Double 38: $60 + 16 = 76$

a) Double 43 _____

b) Double 19 _____

c) Double 26 _____

d) Double 37 _____

e) Double 55 _____

f) Double 73 _____

g) Double 69 _____

2. **Halve** these numbers by halving the tens, halving the units, then combining.

a) Halve 64 _____

b) Halve 36 _____

c) Halve 72 _____

d) Halve 44 _____

e) Halve 54 _____

f) Halve 84 _____

g) Halve 76 _____

h) Halve 92 _____

3.

Shoes	Slippers	Trainers	Sandals	Boots
$56	$42	$65	$24	$48

- Write the cost of two pairs of each type of shoe:

  Shoes _____

  Slippers _____

  Trainers _____

  Sandals _____

  Boots _____

  How much is half the cost of each type of shoe?

  Shoes _____

  Slippers _____

  Trainers _____

  Sandals _____

  Boots _____

4. Fill in this table. The first three rows are done for you.

Half the number	Number	Double the number
12	24	48
120	240	480
1200	2400	4800
	46	
	460	
	4600	
	28	
	280	
	2800	
	34	
	340	
	3400	

5. Use doubling facts to help calculate these sums that are near-doubles:

a) Double 42 = _____   42 + 43 = _____

b) Double 28 = _____   28 + 29 = _____

c) Double 47 = _____   47 + 46 = _____

d) Double 36 = _____   36 + 37 = _____

e) Double 19 = _____   19 + 18 = _____

f) Double 58 = _____   58 + 59 = _____

3E Doubling and halving

Journey length: Dubai–Muscat 450 km

Return journey length: Dubai–Muscat–Dubai 900 km

- Complete the return journey length for these flights:

Plane journey length	Return journey length
340 km	
180 km	
420 km	
270 km	
490 km	
360 km	
Choose 3 journey lengths of your own to calculate: (Use lengths that are multiples of 10.)	

3 Multiplication

Design an aquarium!

- Work with a partner.

Red tail botia $4	Zebra Stripe $7	Polka dot botia $9	Yellow tail botia $5
Red tail zebra $3	Doctor Garra $6	Tiger Botia $8	Dwarf Chain Botia $10

You work at a zoo and are in charge of creating a new aquarium.

You have $1000 to spend on fish for the new aquarium.

- Decide how many of each fish you would like to buy and work out the cost.

- Choose a minimum of 10 fish of each type.

Choose some of each type of fish.

You may want to find more exotic types of fish on the Internet.

Use paper for planning.

- Make a clear list of your final choices, showing the number and cost for each species.

3 Multiplication

- Write an example for each of these.
- Give your questions to a friend to answer.
- Mark your friend's answers and correct any errors.

Write a question involving doubling a 2-digit number:
Write a TU × U multiplication that you solve using the grid method:
Write a question in which you divide a multiple of 100 by 100:
Write a question involving halving a 3-digit number that is a multiple of 10:
Write a multiplication that you solve by partitioning and recombining:
Write a multiplication in which you multiply three single-digit numbers:
Explain how you can solve a × 8 multiplication by doubling:

4 Division

What is division?

Which **operations** is division related to?

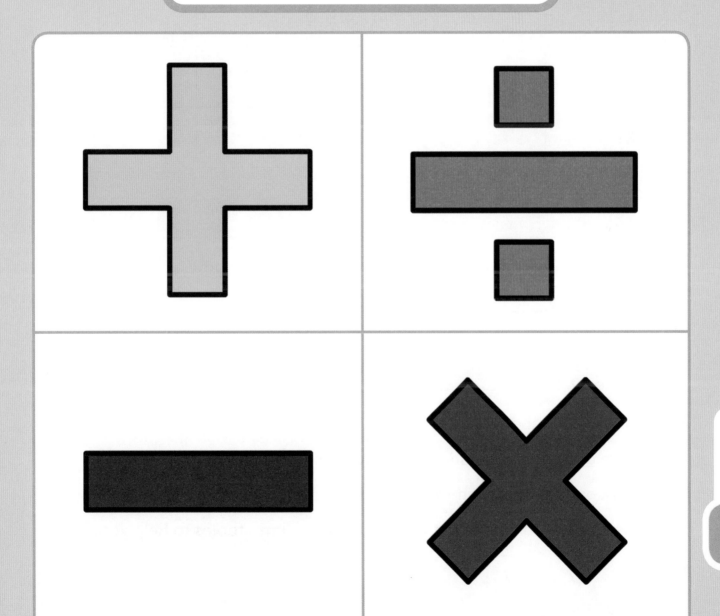

4A Dividing 2-digit numbers by a single-digit number

Discover

I. Investigating **remainders**!

- Complete this table.

Some examples are done for you.

÷	31	32	33	34	35	36	37	38	39	40
÷3	10 r 1									
÷4		8								
÷5			6 r 3							
÷6				5 r 4						

Can you see any patterns? _____

_____

What is the largest remainder that you can have when you:

- divide by 4? _____

- divide by 6? _____

- divide by 9? _____

2. Choose one number from each box.

- Write a division calculation with your two numbers.

- Work out the answer in three jumps or fewer.

- Use a blank number line and knowledge of times tables to help you.

71	67	87
59		79

6	3	4	9

a) I choose: _____

My division is: _____

My working:

0

My answer is: _____

b) I choose: _____

My division is: _____

My working:

0

My answer is: _____

c) I choose: _____

My division is: _____

My working:

0

My answer is: _____

3. When we use 'chunking', we write the calculation vertically.

For example:
What is 93 ÷ 6?

$$93 = 60 + 33$$

$$
\begin{array}{rl}
93 & \\
-\ 60 & \quad 10 \times 6 \\
\hline
33 & \\
-\ 30 & \quad 5 \times 6 \\
\hline
3 & \quad r\ 3
\end{array}
$$

$$93 \div 6 = 15\ r\ 3$$

I need to write the numbers down very carefully so I don't get confused.

- Use the 'chunking' method to solve these calculations:

70 ÷ 6

61 ÷ 4

89 ÷ 4

77 ÷ 4

4A Dividing 2-digit numbers by a single-digit number

Explore

I. Answer these questions.

- Use a method of your choice.

- Do your working on paper.

- Then write your final answer in a sentence.

a) There are 5 chocolate biscuits in a pack.

   You need 80 biscuits for a party.

   How many packs do you need to buy?

   I need to buy _____ packs of biscuits.

b) You have 96 stickers.

   You can put 6 stickers on a page.

   How many pages can you fill?

   I can fill _____ pages.

c) Four children can fit in a canoe.

   How many canoes do you need for 60 children?

   _____

d) There are 95 chairs in a hall.

   The chairs are stored in groups of 5.

   How many groups of chairs are there?

   _____

e) There are 6 eggs in a box.

   A café serves 102 eggs at breakfast.

   How many boxes of eggs does the café use?

   _____

   Did you notice anything that was the same about all these calculations?

   _____

2. Choose a 2-digit number from this section of the 100-square:

81	82	83	84	85	86	87	88	89	90
91	92	93	94	95	96	97	98	99	100

- Divide your number by 2, 3, 4, 5, 6, 9 and 10.

You can use any method. You may be able to do some mentally.

The number I chose was _____

Here are my answers:

Show your working here:

_____ ÷ 2 = _____

_____ ÷ 3 = _____

_____ ÷ 4 = _____

_____ ÷ 5 = _____

_____ ÷ 6 = _____

_____ ÷ 9 = _____

_____ ÷ 10 = _____

How many of your calculations had a remainder? _____

- Use a separate piece of paper to investigate these questions:

Can you find a number where **every** calculation has a remainder?

Can you find a number where **none** of the calculations has a remainder?

4B Rounding answers up or down

I. Work together to write and solve division problems where you need to round **up** the answer.

- Use these in your questions:

  a) people and cars

  b) your own idea

2. Work together to write division problems where you need to round **down** the answer.

- Use these in your questions:

  a) eggs and egg boxes

  b) your own idea

3. Discuss these questions with your partner and work out the answers.

- Write a sentence explaining what you did with the remainder.

    a) A ferry can carry 9 cars.

      How many ferries do you need to take 116 cars across the river?

_____

    b) You share $39 evenly between 2 children.

      How much money does each child receive?

_____

    c) 80 computers are packed in containers.

      Each container holds 6 computers.

      How many full containers are there?

_____

4B Rounding answers up or down

Sea-view restaurant

I. In this restaurant 6 people can sit at each table.

How many tables do the staff need to prepare each day?

	Number of people	Number of tables needed	Show your calculation in this column
Monday	75		
Tuesday	72		
Wednesday	73		
Thursday	67		
Friday	82		

Division

83

2. The restaurant sells the *Sea-view Cookbook*.

One *Cookbook* costs $4.

How many books did the restaurant sell each night?

	Money from cookbook sales	Number of cookbooks sold	Show your calculation in this column
Monday	$56		
Tuesday	$68		
Wednesday	$48		
Thursday	$76		
Friday	0		

Can you suggest a reason why the restaurant did not have any cookbook sales on Friday?

Perhaps they didn't sell any books on Friday because _____

_____

Discover

I. Use the numbers and symbols in this box:

> 2 3 4 5 6 12 15 18 20 24 30
>
> ÷   ×   =

- Make as many correct multiplication and division sentences as you can.

> For example:
>
> $12 \div 3 = 4$ $3 \times 4 = 12$

You can use the numbers and symbols as many times as you like.

You cannot put 2 single-digit numbers together to make a 2-digit number. For example: you can not put 4 and 5 together to make 45.

2. Make up five division calculations that have a remainder of 2.

Challenge yourself. Make them as difficult as you can.

How do you know that the remainder is 2?

_____

_____

4C Multiplication and division as inverse operations

1.

> Here is an example of a division sentence:
>
> $21 \div 7 = 3$
>
> Here are three more number sentences using these numbers:
>
> $21 \div 3 = 7$
>
> $3 \times 7 = 21$
>
> $7 \times 3 = 21$

- Write as many multiplication and division sentences as you can with these numbers:

a) $20 \div$ _____ = _____

c) $54 \div$ _____ = _____

b) $16 \div$ _____ = _____

d) $74 \div$ _____ = _____

How many number sentences did you find? _____

Four number sentences is good.

More than four is **very** good.

Knowing that you found all the number sentences is **excellent**.

2. A friend says: '8 ÷ 2 = 4 so 2 ÷ 8 = 4'.

- Draw a diagram to explain why this is **not** true:

4D Ratio and proportion

Discover

The heights shown for the animals in these pictures are smaller than in real life. Work out the real size for each animal.

Meerkat  7 cm	Gorilla  8 cm	Tiger  9 cm	Sloth  12 cm
This is $\frac{1}{4}$ real **height**.	This is $\frac{1}{20}$ real height.	This is $\frac{1}{10}$ real height.	This is $\frac{1}{4}$ real height.
Real height is _____	Real height is _____	Real height is _____	Real height is _____
Red fox  8 cm	Male lion  9 cm	Armadillo  7 cm	Squirrel 3 cm
This is $\frac{1}{6}$ real height.	This is $\frac{1}{10}$ real height.	This is $\frac{1}{3}$ real height.	This is $\frac{1}{4}$ real height.
Real height is _____	Real height is _____	Real height is _____	Real height is _____
Raccoon  6 cm	Rhesus monkey  10 cm	Brown bear  7 cm	Giant panda  8 cm
This is $\frac{1}{4}$ real height.	This is $\frac{1}{6}$ real height.	This is $\frac{1}{20}$ real height.	This is $\frac{1}{10}$ real height.
Real height is _____	Real height is _____	Real height is _____	Real height is _____

Camel	Rabbit	Ring-tailed lemur	Giraffe
 10 cm	5 cm	 8 cm	5 cm
This is $\frac{1}{20}$ real height.	This is $\frac{1}{5}$ real height.	This is $\frac{1}{5}$ real height.	This is $\frac{1}{100}$ real height.
Real height is _____	Real height is _____	Real height is _____	Real height is _____

Which of these animals is the tallest in real life? _____

Which of these animals is the shortest in real life? _____

Which two pairs of these animals have similar heights in real life?

_____

- Write two more questions from this information:

_____

_____

- Write two animals whose height you do not know.

    Find out their heights and fill in the boxes:

Picture:	Picture:
The real height of a _____ is _____ cm. I need to divide this by _____ to put a picture in the box.	The real height of a _____ is _____ cm. I need to divide this by _____ to put a picture in the box.

4D Ratio and proportion

Explore

Recipes for one person

Here are some recipes for a meal of lentil soup, butter chicken curry and chocolate ice-cream.

The recipes are for different numbers of people.

- Change each recipe so that the amounts are correct for **one** person.

Spicy Lentil Soup for 2 people	Spicy Lentil Soup for 1 person
1 onion, chopped	
2 large carrots	
150 g red lentils	
1 litre vegetable stock	
1 lime	

cumin, ginger and chilli flakes to taste	cumin, ginger and chilli flakes to taste
coriander leaves to decorate	coriander leaves to decorate

Butter Chicken Curry for 4 people

200 g butter

1 large onion, chopped

4 teaspoons curry powder

4 chicken breast fillets, cubed

6 fresh tomatoes,
peeled and chopped

150 ml tinned tomatoes

Butter Chicken Curry for 1 person

Chocolate Ice-cream for 6 people

120 g dark chocolate, in pieces

300 ml milk

90 g sugar

3 egg yolks

300 ml cream

Chocolate Ice-cream for 1 person

4 Division

The answer is 6.

- Make up 10 different division calculations with this answer.

An extra challenge! You must use these words somewhere in your questions:

(Hint – Cross each word out as you use it!)

share	equally	owls	divide	dollars	quotient

pencils   divided by   camels   groups   remainder   each

ducks   stickers

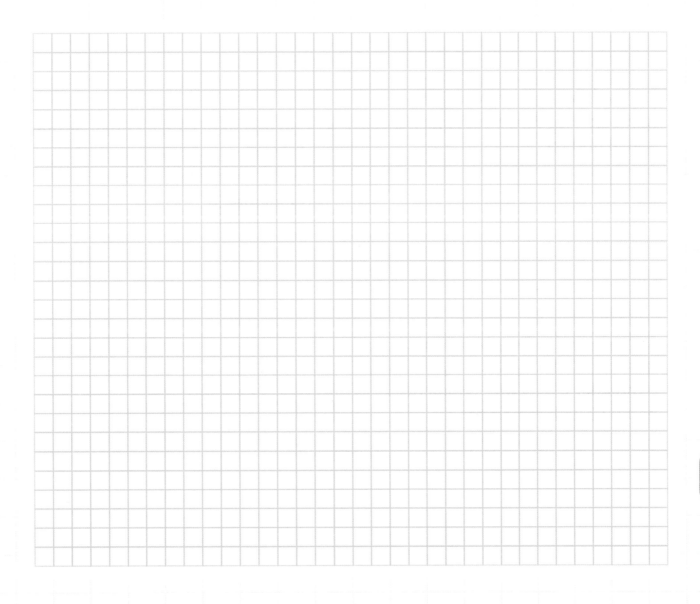

4 Division

÷2? ÷5? ÷10? ÷4? ÷6?

- Show the division calculations for each of these 2-digit number puzzles.

1. Can you find one or more 2-digit numbers that divide exactly by 2, 5 and 10?

2. Can you find one or more 2-digit numbers that divide exactly by 2, 3, 4, 6, 8 and 12?

3. Can you find a 2-digit number less than 30 that divides by 2 with remainder 1, by 5 with remainder 2 and by 6 with remainder 3?

4. Can you find a 2-digit number that divides by 4 with remainder 1, by 5 with remainder 1 and by 6 with remainder 1?

5. Can you find a 2-digit number that divides by 4 with remainder 3, by 5 with remainder 3 and by 6 with remainder 3?

6. Make up a similar puzzle of your own:

5 Fractions

What fractions can you see?

When do we use fractions in everyday life?

5A Ordering and comparing fractions

Discover

1. In this diagram you can see $\frac{1}{5}$ of a whole.

    The red part is the shaded fraction.

    The white part is the unshaded fraction.

    $\frac{4}{5}$ of the whole is unshaded.

*   Draw diagrams to illustrate these in a similar way:

    a) You can see $\frac{1}{4}$ of a whole.

    What fraction is unshaded?

    ___$\frac{3}{4}$___ is unshaded.

    b) You can see $\frac{7}{8}$ of a whole

    What fraction is unshaded?

    ___$\frac{1}{8}$___ is unshaded.

    c) You can see $\frac{5}{10}$ of a whole.

    What fraction is unshaded?

    ___$\frac{5}{10}$___ is unshaded.

    d) You can see $\frac{1}{3}$ of a whole.

    What fraction is unshaded?

    ___$\frac{2}{3}$___ is unshaded.

2. There are 10 cubes in a bag.

You take out one cube.

What fraction of the cubes is in your hand?

This is $\frac{1}{10}$ of the cubes.

What fraction is left in the bag?

$\frac{9}{10}$ are left in the bag.

- Make up three more questions like this:

   a) There are _____ 8 _____ cubes in a bag. I take out _____ 3 _____ cubes.

   This is _____ $\frac{3}{8}$ _____ of the cubes _____ $\frac{5}{8}$ _____ are left in the bag.

   b) There are _____ 100 _____ cubes in a bag. I take out _____ 75 _____ cubes.

   This is _____ $\frac{15}{100}$ _____ of the cubes _____ $\frac{25}{100}$ _____ are left in the bag.

   c) There are _____ 5 _____ cubes in a bag. I take out _____ 4 _____ cubes.

   This is _____ $\frac{4}{5}$ _____ of the cubes _____ $\frac{1}{5}$ _____ are left in the bag.

3. You and your friend share a bar of chocolate.
The bar of chocolate has 8 pieces.

You and your friend eat it all. You eat $\frac{3}{8}$ of the chocolate bar.

What fraction does your friend eat? _____ $\frac{5}{8}$ _____

Using **eighths**, work out all the different ways you can share the chocolate.
Write them clearly:

$\frac{6}{8}$ and $\frac{2}{8}$ $\frac{1}{8}$ and $\frac{7}{8}$

Did you include one way where you were very greedy?

5A Ordering and comparing fractions

Explore

$$\frac{1}{3} < \frac{3}{5}$$

1. Write a fraction in each space to make the statement true:

a) $\frac{1}{4} < \underline{\frac{3}{4}} < 1$

b) $\frac{7}{8} > \underline{\frac{4}{8}} > \frac{1}{8}$

c) $\frac{2}{5} < \underline{\frac{3}{5}} < \frac{4}{5}$

d) $\frac{7}{10} > \underline{\frac{5}{10}} > \frac{3}{10}$

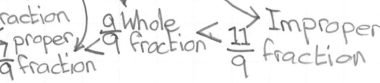

$\frac{0}{9}$ zero fraction

proper fraction

$\frac{7}{9}$ proper fraction

a whole $\frac{9}{9}$ fraction

$\frac{11}{9}$ Improper fraction

2. Use a ruler to divide these squares into quarters.

   Make each one a different pattern.

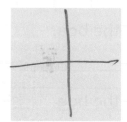  

3.

> Look at this diagram and the number sentence:
>
> | | | |
> |---|---|---|
>
> $\frac{1}{3} + \frac{1}{3} + \frac{1}{3} = \frac{3}{3} = 1$

• Complete these number sentences in the same way.

a)

$\frac{1}{4} + \underline{\frac{1}{4} + \frac{1}{4} + \frac{1}{4}} = \frac{4}{4} = 1$

b)

$\frac{1}{8} + \underline{\frac{1}{8}\ \frac{1}{8}\ \frac{1}{8}\ \frac{1}{8}\ \frac{1}{8}\ \frac{1}{8}\ \frac{1}{8}} = \underline{\frac{8}{8}} = \underline{1}$

c)

_____ = _____ = _____

d)

_____ = _____ = _____

4. What fraction of each pizza has been eaten?

What fraction is left?

- Complete the table.

   The first has been done as an example.

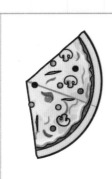	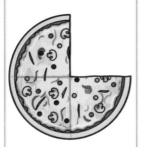		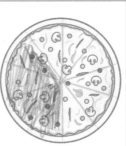	
			You choose how much is eaten.	You choose how many pieces and how much is eaten.
Eaten $\frac{3}{5}$	Eaten $\frac{1}{4}$	Eaten $\frac{5}{8}$	Eaten $\frac{4}{10}$	Eaten $\frac{1}{2}$
Not eaten $\frac{2}{5}$	Not eaten $\frac{3}{4}$	Not eaten $\frac{3}{8}$	Not eaten $\frac{6}{10}$	Not eaten $\frac{1}{2}$

5B Equivalent fractions

Discover

I. • Fold a sheet of paper in half.

 • Colour $\frac{1}{2}$.

 • Fold again to make four equal pieces.

   How many quarters are coloured? ___$\frac{2}{4}$___

 • Fold again to make eight equal pieces.

   How many eighths are coloured?

   ___$\frac{4}{8}$___

 • Complete:

   $$\frac{1}{2} = \frac{2}{4} = \frac{4}{8}$$

   These are **equivalent fractions**.

2. Fill in the fractions on the wall.

 You can see the fractions that are the same size on the fraction wall.

1			
$\frac{1}{2}$		$\frac{1}{2}$	

3. Use a set of number cards I–10.

 • Work with a partner to find pairs of equivalent fractions.

   For example: $\frac{1}{2} = \frac{4}{8}$

 • Record your answers:

   $\frac{2}{5} = \frac{4}{10}$ $\frac{2}{4} = \frac{4}{8}$

4. Here is part of the 4× table below the 1× table:

× 1	1	2	3	4	5	6
× 4	4	8	12	16	20	24

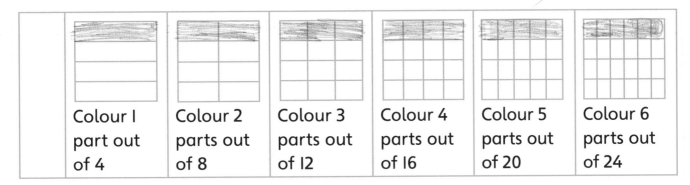

Colour 1 part out of 4	Colour 2 parts out of 8	Colour 3 parts out of 12	Colour 4 parts out of 16	Colour 5 parts out of 20	Colour 6 parts out of 24

What do you notice about the fraction coloured in each column?

All the fractions are equivalent to each other.

- Write some fractions that are equivalent to $\frac{1}{4}$:

  $\frac{2}{8}$, $\frac{4}{16}$, $\frac{8}{32}$, $\frac{16}{64}$

5. Here is part of the 2× table below the 1× table:

× 1	1	2	3	4	5	6	7	8	9	10
× 2	2	4	6	8	10	12	14	16	18	20

What do you notice about each pair of numbers?

That the top number is the half of the bottom number.

- Read them as fractions.

  What do you notice?

  Their all equivalent

5B Equivalent fractions

1. Name each fraction and match equivalent fractions.

   One has been done for you: $\frac{1}{2}$ is equivalent to $\frac{4}{8}$.

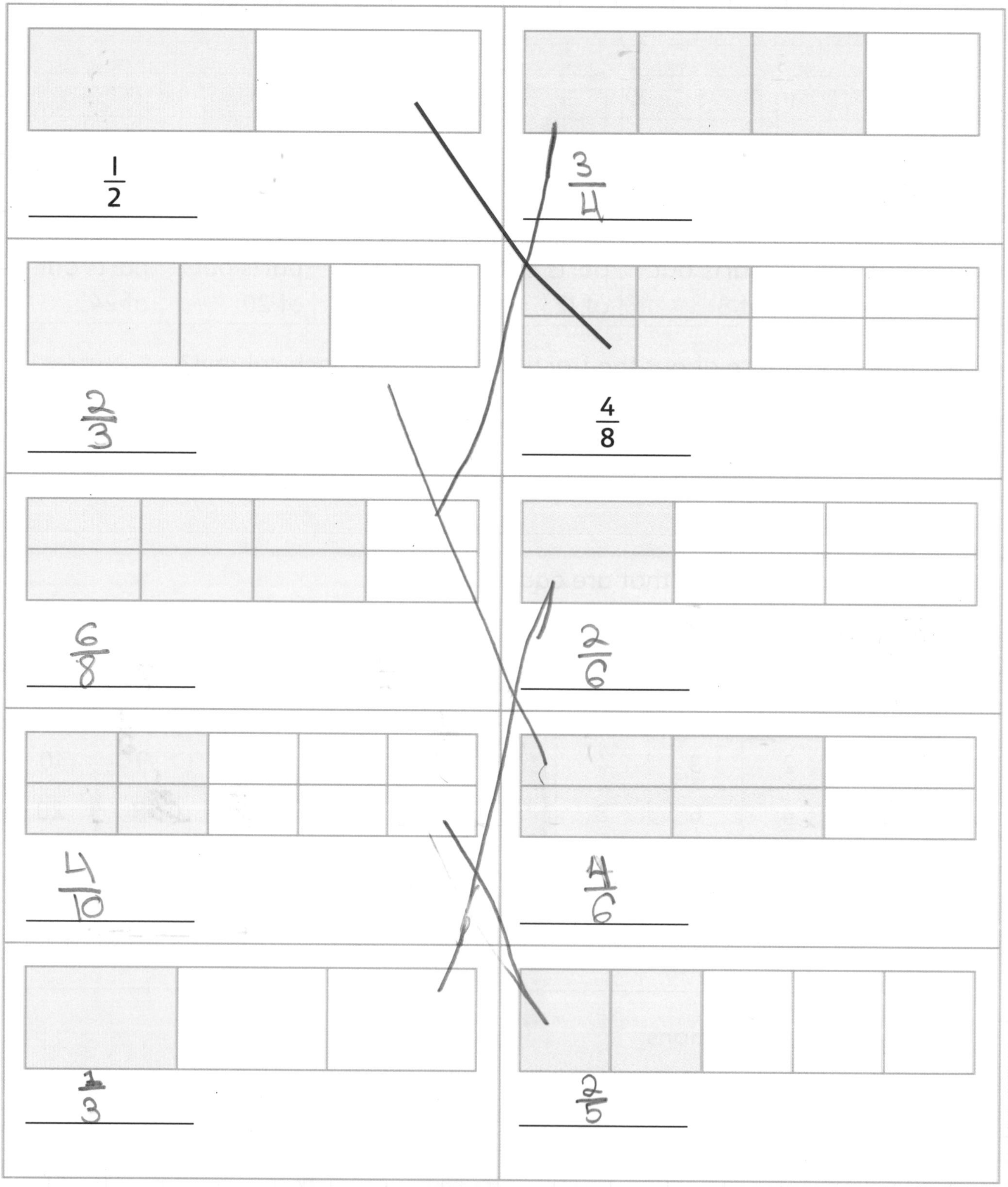

$\frac{1}{2}$

$\frac{3}{4}$

$\frac{2}{3}$

$\frac{4}{8}$

$\frac{6}{8}$

$\frac{2}{6}$

$\frac{4}{10}$

$\frac{4}{6}$

$\frac{1}{3}$

$\frac{2}{5}$

2. Circle the odd one out.

For example: $\frac{1}{5}$ $\left(\frac{3}{8}\right)$ $\frac{2}{10}$

a) $\frac{1}{2}$ $\frac{4}{8}$ $\left(\frac{2}{5}\right)$

b) $\frac{6}{8}$ $\frac{3}{4}$ $\left(\frac{4}{10}\right)$

c) $\frac{6}{10}$ $\left(\frac{3}{4}\right)$ $\frac{3}{5}$

d) $\frac{2}{8}$ $\left(\frac{4}{10}\right)$ $\frac{1}{4}$

e) $\frac{2}{3}$ $\frac{4}{6}$ $\left(\frac{1}{2}\right)$

f) $\frac{4}{10}$ $\frac{2}{5}$ $\left(\frac{1}{2}\right)$

g) $\frac{4}{5}$ $\left(\frac{7}{8}\right)$ $\frac{8}{10}$

h) $\left(\frac{2}{3}\right)$ $\frac{3}{5}$ $\frac{6}{10}$

3. Choose a number of cubes to make a shape that is $\frac{3}{4}$ one colour, $\frac{1}{4}$ another colour.

• Sketch your shape.

• Write the number of cubes used.

Sketch 1	Sketch 2	Sketch 3
Number of cubes used	Number of cubes used	Number of cubes used
100	64	72

Can you make a $\frac{3}{4}$, $\frac{1}{4}$ pattern with 10 cubes? _____

What do you notice about the numbers that you have used?

_____

Make predictions. Can you make a $\frac{3}{4}$, $\frac{1}{4}$ pattern with these?

Circle the correct answer.

    a) 24 red cubes and 8 yellow cubes

       Yes/ No

    b) 30 blue cubes and 10 pink cubes

       Yes/ No

    c) 19 green cubes and 7 red cubes

       Yes/ No

    d) 120 white cubes and 40 yellow cubes

       Yes/ No

Add two more examples that you can use:

e) _____

f) _____

5C Using equivalence to order fractions

Discover

I. Write the equivalent fractions in the correct place on the number line:

a)

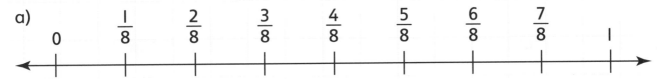

$$\frac{3}{4}, \frac{1}{8}, \frac{1}{2}, \frac{7}{8}, \frac{1}{4}, \frac{5}{8}, \frac{3}{8}$$

b)

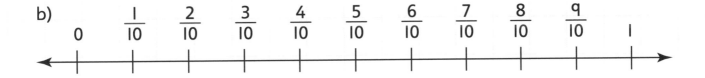

$$\frac{3}{5}, \frac{1}{10}, \frac{1}{2}, \frac{4}{5}, \frac{3}{10}, \frac{7}{10}, \frac{2}{5}, \frac{9}{10}, \frac{1}{5}$$

c)

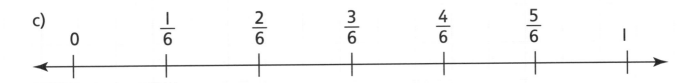

$$\frac{1}{2}, \frac{1}{6}, \frac{5}{6}, \frac{2}{3}, \frac{1}{3}$$

2. Write at least six fractions in each part of this table:

Less than (<) $\frac{1}{2}$	Greater than (>) $\frac{1}{2}$
$\frac{2}{6}$ $\frac{4}{11}$ $\frac{3}{10}$ $\frac{5}{12}$ $\frac{1}{3}$ $\frac{1}{8}$	$\frac{4}{6}$ $\frac{7}{11}$ $\frac{8}{10}$ $\frac{7}{12}$ $\frac{2}{3}$ $\frac{7}{8}$

3. Here is a new fraction wall. Look at it carefully.

What fractions does it show? Discuss with your partner.

- Fill in all the fractions that you can:

1											
$\frac{1}{2}$						$\frac{1}{2}$					
$\frac{1}{4}$			$\frac{1}{4}$			$\frac{1}{4}$			$\frac{1}{4}$		
$\frac{1}{6}$		$\frac{1}{6}$		$\frac{1}{6}$		$\frac{1}{6}$		$\frac{1}{6}$		$\frac{1}{6}$	
$\frac{1}{12}$	$\frac{1}{12}$	$\frac{1}{12}$	$\frac{1}{12}$	$\frac{1}{12}$	$\frac{1}{12}$	$\frac{1}{12}$	$\frac{1}{12}$	$\frac{1}{12}$	$\frac{1}{12}$	$\frac{1}{12}$	$\frac{1}{12}$

- Write pairs of equivalent fractions that you can see on this wall:

$\frac{1}{2} = \frac{2}{4}$ $\frac{3}{6} = \frac{6}{12}$

$1 = \frac{12}{12}$ $1 = \frac{4}{4}$

5C Using equivalence to order fractions

Explore

1. Use this fraction wall to find some fractions between $\frac{1}{4}$ and $\frac{1}{2}$:

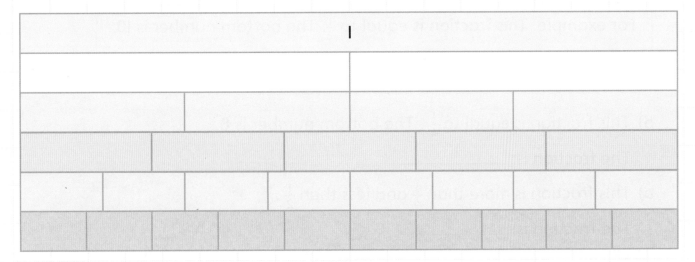

2. Write an equivalent fraction for each of these:

a) $\frac{1}{5} =$ _____

b) $\frac{1}{2} =$ _____

c) $\frac{3}{5} =$ _____

d) $\frac{2}{8} =$ _____

e) $\frac{3}{4} =$ _____

f) $\frac{4}{10} =$ _____

3. Are these statements true or false? Circle the correct answer.

For example:
$\frac{1}{2}$ is less than $\frac{2}{3}$.   ／ False

a) Three-quarters is equal to six-eighths.

   True / False

b) $\frac{6}{10} > \frac{1}{2}$.

   True / False

c) $\frac{1}{4}$ is the same as $\frac{3}{8}$.

   True / False

d) $\frac{3}{10} < \frac{3}{8}$.

   True / False

e) $\frac{2}{5}$ is equal to $\frac{4}{10}$.

   True / False

4. Work with a partner to find these fractions.

> For example: This fraction is equal to $\frac{1}{2}$. The bottom number is 10.
>
> The fraction is: $\frac{5}{10}$

a) This fraction is equal to $\frac{1}{4}$. The bottom number is 8.

The fraction is: _____

b) This fraction is more than $\frac{1}{4}$ and less than $\frac{1}{2}$.

The fraction is: _____

c) This fraction is equal to $\frac{1}{2}$.

The top number and bottom number have a total of 12.

The fraction is: _____

d) The top number in this fraction is 2 less than the bottom number.

The fraction is equal to $\frac{3}{4}$.

The fraction is: _____

e) The top number and bottom number are both odd numbers and have a total of 6.

The fraction is: _____

5. Now make up similar sentences to describe these fractions:

a) $\frac{9}{10}$ _____

_____

b) $\frac{1}{3}$ _____

_____

c) $\frac{7}{8}$ _____

_____

5D Mixed numbers

Discover

1. Work out, draw and write the **mixed number**.

> For example: When 5 bricks make one tower, how many towers do 14 bricks make?
>
> 14 bricks make $2\frac{4}{5}$ towers

a) 8 bricks make one tower. How many towers do 14 bricks make?

b) 4 bricks make one tower. How many towers do 14 bricks make?

c) Now choose your own tower height and draw the result:

2. Here are the ages in years of some groups of friends.

   Put each group in order from the youngest to the oldest:

$6\frac{1}{4}$ years $\qquad$ $9\frac{2}{3}$ years $\qquad$ $8\frac{3}{4}$ years $\qquad$ $7\frac{1}{6}$ years $\qquad$ $8\frac{1}{4}$ years	_____ < _____ < _____ < _____ < _____
$8\frac{1}{3}$ years $\qquad$ $9\frac{1}{3}$ years $\qquad$ $8\frac{1}{6}$ years $\qquad$ 9 years $\qquad$ $8\frac{1}{2}$ years	_____ < _____ < _____ < _____ < _____
$7\frac{2}{3}$ years $\qquad$ $7\frac{3}{4}$ years $\qquad$ $7\frac{1}{4}$ years $\qquad$ $7\frac{1}{2}$ years $\qquad$ 7 years	_____ < _____ < _____ < _____ < _____

3. Draw the fraction and mixed number on the number line.
Circle the fraction that is closer to I.

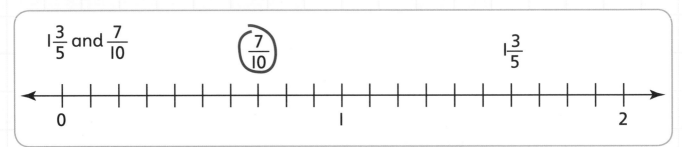

$1\frac{3}{5}$ and $\frac{7}{10}$

a) $1\frac{1}{4}$ and $\frac{5}{8}$

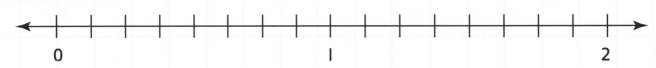

b) $\frac{3}{4}$ and $1\frac{3}{8}$

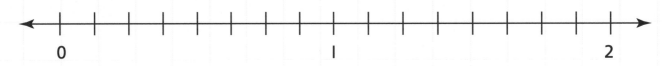

c) $1\frac{2}{5}$ and $\frac{9}{10}$

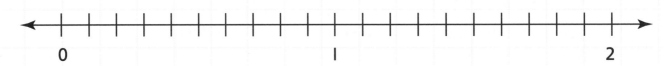

d) $\frac{7}{8}$ and $1\frac{1}{2}$

e) $\frac{8}{10}$ and $1\frac{1}{5}$

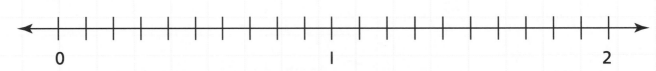

5D Mixed numbers

I. It takes $\frac{1}{4}$ of an hour to cook a large tray of biscuits.

Only one tray can fit in the oven at a time.

How many hours does it take to cook these?

a) 6 trays of biscuits take _____ hours to cook.

b) 9 trays of biscuits take _____ hours to cook.

c) 11 trays of biscuits take _____ hours to cook.

d) 21 trays of biscuits take _____ hours to cook.

e) 17 trays of biscuits take _____ hours to cook.

f) 45 trays of biscuits take _____ hours to cook.

2. At the beginning of the day the bakery has 10 of each type of cake.

At the end of the day the baker checks how much of each type of cake is left.

• Work out how much of each cake has been sold.

Amount of cake left (out of 10)	Amount left	Amount sold
	$4\frac{3}{4}$	$5\frac{1}{4}$

5 Fractions

My Week

How much of each day do you spend at school?

How much of each day do you spend doing homework?

How much of each day do you spend sleeping?

How much of each day do you spend watching television?

Monday			
Sleeping	Sleeping	Sleeping	Sleeping
Sleeping	Sleeping	Sleeping	Watching television
Travelling to school	School	School	School
Playing	School	School	School
Playing	Homework	Watching television	Reading
Reading	Sleeping	Sleeping	Sleeping

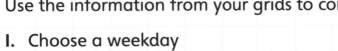

I spend $\frac{10}{24}$ of my day sleeping.

I spend $\frac{6}{24}$ of my day at school–that is the same as $\frac{1}{4}$.

Use the information from your grids to complete these tables.

I. Choose a weekday

Activity	Number of hours	Fraction of the day

2. Choose a day at the weekend.

Activity	Number of hours	Fraction of the day

Choose one of the grids and use the information to complete the chart on the right.

Use the information from your grids and tables to write four sentences about the way you spend your day.

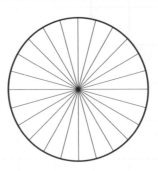

114

I. _____

2. _____

3. _____

4. _____

5 Fractions

- Use the pattern of lines to divide the square in different ways.

  Use the same shape for each of the fractions in one square.

Divide into $\frac{3}{3}$.
Is there more than one way to do this?

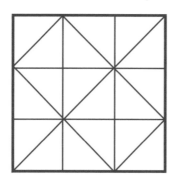

Divide into $\frac{6}{6}$.
Is there more than one way to do this?

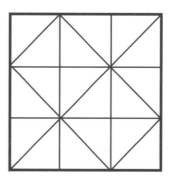

Divide into $\frac{9}{9}$.

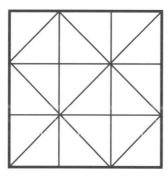

Use different colours to show ways of dividing the square in half.
How many ways can you find?

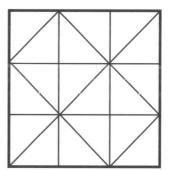

You need to use more lines to divide the square into four quarters ($\frac{4}{4}$).

What is the smallest number of lines you can use to do this? The quarters must all be the same shape.

_____

Draw your final answer in the square on the right:

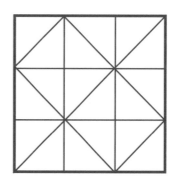

6 Decimals and Fractions

How and why do we use **decimals**?

6A Decimals and tenths

Discover

1. You need: Set of **tenths** cards ($\frac{1}{10}$, $\frac{2}{10}$, $\frac{3}{10}$, $\frac{4}{10}$, $\frac{5}{10}$, $\frac{6}{10}$, $\frac{7}{10}$, $\frac{8}{10}$, $\frac{9}{10}$, 1)

   Set of **decimal fraction** cards (0.1, 0.2, 0.3, 0.4, 0.5, 0.6, 0.7, 0.8, 0.9, 1.0)

   Set of **fifths** cards ($\frac{1}{5}$, $\frac{2}{5}$, $\frac{3}{5}$, $\frac{4}{5}$, $\frac{5}{5}$)

   - Spread the cards face down in three piles.

   - Choose:

     - two cards from the tenths set

     - two from the decimal fractions set

     - one from the fifths set.

   - Put the numbers in order, from smallest to largest.

   - Record your order using the **less than** sign (<):

   _____ < _____ < _____ < _____ < _____

   - Repeat with five more cards.

   - Continue to use all the cards.

   - Record your ordered numbers here:

   _____ < _____ < _____ < _____ < _____

   _____ < _____ < _____ < _____ < _____

   _____ < _____ < _____ < _____ < _____

   _____ < _____ < _____ < _____ < _____

   _____ < _____ < _____ < _____ < _____

2. Look at these bags of oranges.

Each customer wants to buy exactly 5 kg.

- Draw lines to match pairs that total 5 kg.

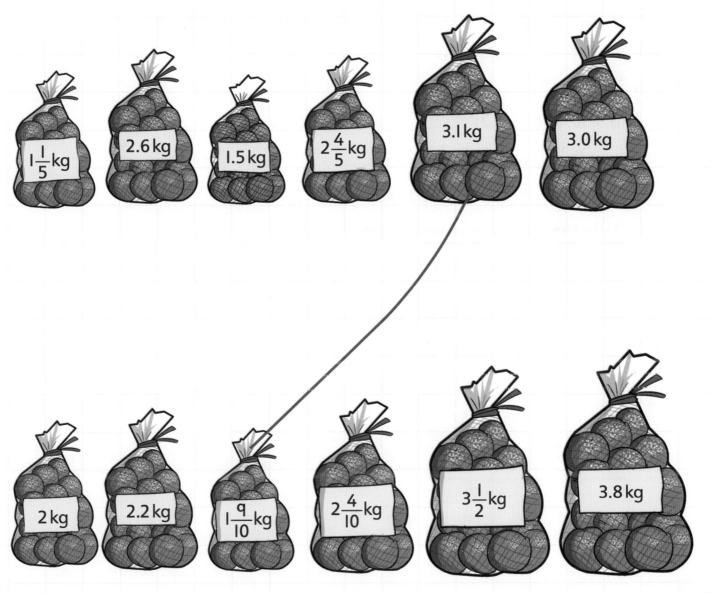

6A Decimals and tenths

1. Complete these decimal number sequences.
   The first one has been done for you.

   a) 0.6, 0.7, 0.8, _____0.9_____, _____1.0_____, _____1.1_____, 1.2, 1.3

   b) 1. 7, 1.8, _____, _____, _____, _____, 2.3, 2.4

   c) 1.4, 1.3, 1.2, _____, _____, _____, 0.8, 0.7

   d) 5.3, 5.2, 5.1, _____, _____, _____, _____, 4.6

   e) 3.5, 3.6, 3.7, _____, _____, _____, _____

   f) 10.5, 10.4, 10.3, 10.2, _____, _____, _____, _____

2. Look at these four designs for flags.

   How much of each is coloured yellow?

   • Write the answer in tenths and as a decimal.

   One is shown as an example.

$\frac{2}{10}$	0.2					

- Design your own flags. Use two colours.

  The colours I chose were _____ and _____.

- Fill in the colours you choose.

  Write the first colour in tenths and the second as a decimal fraction of each design.

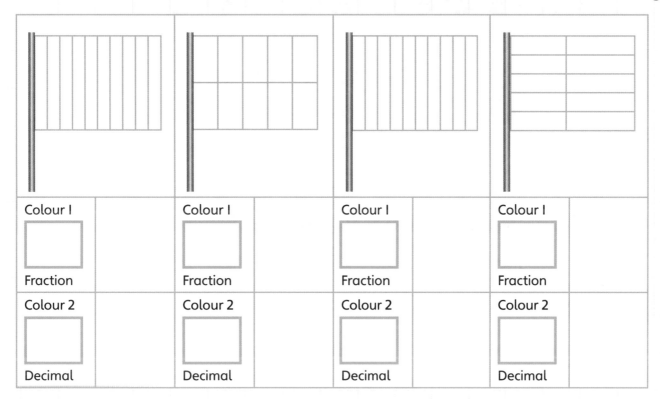

Colour 1		Colour 1		Colour 1		Colour 1	
Fraction		Fraction		Fraction		Fraction	
Colour 2		Colour 2		Colour 2		Colour 2	
Decimal		Decimal		Decimal		Decimal	

3. Put these in order, from smallest to largest.

> For example: $1\frac{1}{5} < 1\frac{3}{10} < 1.5 < 1.8$

a) 2.4 $2\frac{3}{5}$ $2\frac{7}{10}$ 2.1 _____

b) $5\frac{4}{5}$ 6   5.3 $5\frac{9}{10}$ _____

c) 4.9 $4\frac{2}{5}$ $4\frac{6}{10}$ 4.2 _____

d) $3\frac{1}{10}$ 3.7 $3\frac{1}{5}$ $3\frac{4}{10}$ _____

6B Using decimals for tenths and hundredths

Discover

1. Use this '100 cents square' to help you complete the table:

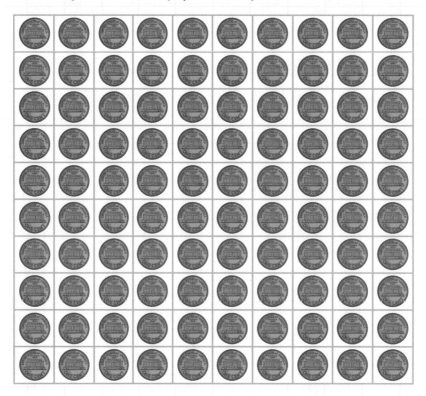

- Write each fraction as a number of cents and in dollars.

Fraction	Number of cents	Amount in dollars
$\frac{1}{2}$	50	$0.50
$\frac{1}{10}$		
$\frac{1}{4}$		
$\frac{1}{5}$		
$\frac{3}{4}$		
$\frac{3}{10}$		
$\frac{3}{5}$		
$\frac{1}{100}$		

2. You have a bag of money containing dollar notes ($1), quarters (25¢) and cents (1¢).

You can only take 5 coins or notes out of the bag.

What different amounts can you make?

There are quite a lot of possibilities!

What is the smallest possible amount? _____

What is the largest possible amount? _____

6B Using decimals for tenths and hundredths

1. Change these distances from centimetres to metres.

> For example: 125 cm ⟶ 1.25 m

a) 178 cm ⟶

b) 300 cm ⟶

c) 432 cm ⟶

d) 10 cm ⟶

e) 4 cm ⟶

f) 502 cm ⟶

2. How much is in each purse?

- Write your answer in dollars.

$1.26

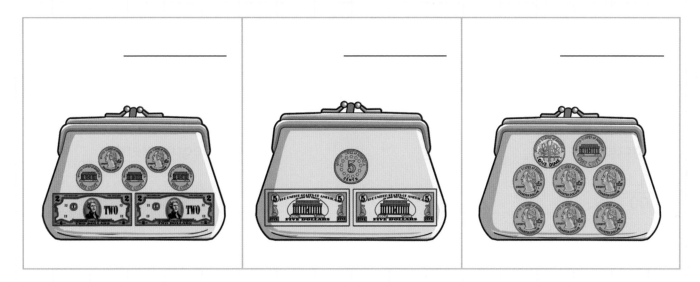

_____ _____ _____

3. Measure these lines.

- Record the lengths in millimetres and centimetres.

  An example is shown.

	Length (mm)	Length (cm)
——	12 mm	1.2 cm
————		
————————		
——————		
—————		
———		
———————		

Discover

I. Play some fraction and decimal matching games with a set of cards.

Try to learn the matching pairs.

- Match each **decimal equivalent** with the correct fraction.
  The first one has been done for you.

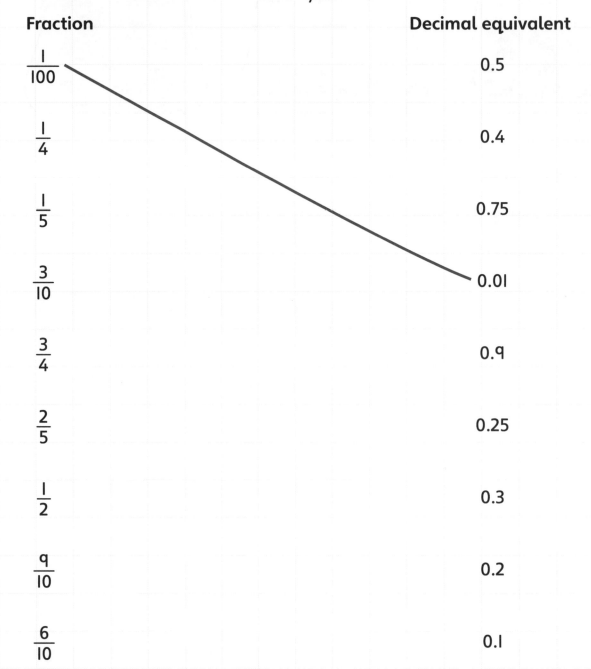

Fraction	Decimal equivalent
$\frac{1}{100}$	0.5
$\frac{1}{4}$	0.4
$\frac{1}{5}$	0.75
$\frac{3}{10}$	0.01
$\frac{3}{4}$	0.9
$\frac{2}{5}$	0.25
$\frac{1}{2}$	0.3
$\frac{9}{10}$	0.2
$\frac{6}{10}$	0.1
$\frac{1}{10}$	0.6

2. Complete this table using your own fractions.

Fraction		Decimal equivalent
	=	
	=	
	=	
	=	
	=	
	=	
	=	

6C Equivalent fractions and decimals

I. Use a strip of paper divided into ten squares to help you complete this table.

Words	Decimal	Fraction	Equivalent fraction	Number line
one tenth				0 ———————— 1
	0.2			0 ———————— 1
		$\frac{3}{10}$		0 ———————— 1
			$\frac{2}{5}$	0 ———————— 1
				0 ———— $\frac{1}{2}$ ———— 1
six tenths				0 ———————— 1
	0.7			0 ———————— 1
		$\frac{8}{10}$		0 ———————— 1
				0 ———— $\frac{9}{10}$ 1
ten tenths	1.0	1	1	0 ———————— 1

2. Colour in the blank 100-squares below to find the decimal equivalent fractions.

For example: $\frac{20}{100} = \frac{2}{10} = 0.2$

1	2	3	4	5	6	7	8	9	10
11	12	13	14	15	16	17	18	19	20
21	22	23	24	25	26	27	28	29	30
31	32	33	34	35	36	37	38	39	40
41	42	43	44	45	46	47	48	49	50
51	52	53	54	55	56	57	58	59	60
61	62	63	64	65	66	67	68	69	70
71	72	73	74	75	76	77	78	79	80
81	82	83	84	85	86	87	88	89	90
91	92	93	94	95	96	97	98	99	100

c) $\frac{3}{4}$

1	2	3	4	5	6	7	8	9	10
11	12	13	14	15	16	17	18	19	20
21	22	23	24	25	26	27	28	29	30
31	32	33	34	35	36	37	38	39	40
41	42	43	44	45	46	47	48	49	50
51	52	53	54	55	56	57	58	59	60
61	62	63	64	65	66	67	68	69	70
71	72	73	74	75	76	77	78	79	80
81	82	83	84	85	86	87	88	89	90
91	92	93	94	95	96	97	98	99	100

a) $\frac{1}{4}$

1	2	3	4	5	6	7	8	9	10
11	12	13	14	15	16	17	18	19	20
21	22	23	24	25	26	27	28	29	30
31	32	33	34	35	36	37	38	39	40
41	42	43	44	45	46	47	48	49	50
51	52	53	54	55	56	57	58	59	60
61	62	63	64	65	66	67	68	69	70
71	72	73	74	75	76	77	78	79	80
81	82	83	84	85	86	87	88	89	90
91	92	93	94	95	96	97	98	99	100

d) $\frac{3}{10}$

1	2	3	4	5	6	7	8	9	10
11	12	13	14	15	16	17	18	19	20
21	22	23	24	25	26	27	28	29	30
31	32	33	34	35	36	37	38	39	40
41	42	43	44	45	46	47	48	49	50
51	52	53	54	55	56	57	58	59	60
61	62	63	64	65	66	67	68	69	70
71	72	73	74	75	76	77	78	79	80
81	82	83	84	85	86	87	88	89	90
91	92	93	94	95	96	97	98	99	100

b) $\frac{1}{2}$

1	2	3	4	5	6	7	8	9	10
11	12	13	14	15	16	17	18	19	20
21	22	23	24	25	26	27	28	29	30
31	32	33	34	35	36	37	38	39	40
41	42	43	44	45	46	47	48	49	50
51	52	53	54	55	56	57	58	59	60
61	62	63	64	65	66	67	68	69	70
71	72	73	74	75	76	77	78	79	80
81	82	83	84	85	86	87	88	89	90
91	92	93	94	95	96	97	98	99	100

e) $\frac{7}{100}$

1	2	3	4	5	6	7	8	9	10
11	12	13	14	15	16	17	18	19	20
21	22	23	24	25	26	27	28	29	30
31	32	33	34	35	36	37	38	39	40
41	42	43	44	45	46	47	48	49	50
51	52	53	54	55	56	57	58	59	60
61	62	63	64	65	66	67	68	69	70
71	72	73	74	75	76	77	78	79	80
81	82	83	84	85	86	87	88	89	90
91	92	93	94	95	96	97	98	99	100

6D Finding fractions of shapes and numbers

Discover

1.

> Here are 6 cubes:
>
>
>
> 2 equal parts – halves   3 equal parts – thirds   6 equal parts – sixths

Now try it with 12 cubes.

How many ways can you divide the cubes to give equal amounts?

- Draw your answers and write the fraction.

> For example:
>
>
>
> 2 equal parts – $\frac{1}{2}$s

2. How do you find a **half**?

To find half I _____

_____

How do you find a **third**?

To find a third I _____

_____

How do you find a **quarter**?

To find a quarter I _____

_____

3. Look at this diagram:

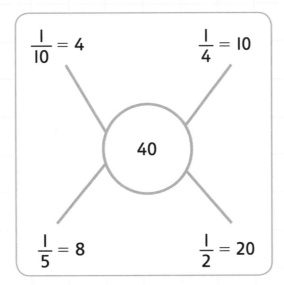

$$\frac{1}{10} = 4 \qquad\qquad \frac{1}{4} = 10$$

$$40$$

$$\frac{1}{5} = 8 \qquad\qquad \frac{1}{2} = 20$$

Can you add any more fractions of 40 that give whole number answers?

_____

- Draw similar diagrams for two of these numbers: 30, 20, 24, 50.

Do you think you found all the answers? _____

How can you be sure? _____

_____

6D Finding fractions of shapes and numbers

How do you find $\frac{3}{4}$ of 20?

First find $\frac{1}{4}$ of 20 by dividing by 4. $\frac{1}{4}$ of 20 = 5

Then multiply by 3 to give $\frac{3}{4}$. $\frac{3}{4}$ of 20 = 3 × 5 = 15

- Find $\frac{3}{4}$ of each of these:

a) 12

b) 40

c) 28

2. How do you think you find $\frac{2}{3}$ of a number?

- First _____

- Then _____

- Find $\frac{2}{3}$ of:

a) 15

b) 24

c) 66

3. Work out the values of the fractions.

Put $<$, $>$ or $=$ in each statement to make it correct.

For example: $\frac{2}{3}$ of 30 $\underline{\quad > \quad}$ $\frac{3}{4}$ of 24

$\frac{1}{3}$ of 30 is 10 so $\frac{2}{3}$ is 20

$\frac{1}{4}$ of 24 is 6 so $\frac{3}{4}$ is 18

a) $\frac{1}{2}$ of 90 $\underline{\qquad}$ $\frac{2}{3}$ of 39	
b) $\frac{3}{4}$ of 40 $\underline{\qquad}$ $\frac{1}{2}$ of 58	
c) $\frac{2}{3}$ of 69 $\underline{\qquad}$ $\frac{3}{4}$ of 80	
d) $\frac{1}{2}$ of 38 $\underline{\qquad}$ $\frac{2}{3}$ of 33	

6 Decimals and fractions

The students in your class decide to go to a pizza restaurant for a meal together.

The number of students in my class is

_____.

1. How many pizzas are needed for:

   a) $\frac{1}{4}$ pizza each? _____

   b) $\frac{1}{3}$ pizza each? _____

   c) $\frac{1}{2}$ pizza each? _____

2. You all decide to have a 0.25 litre milkshake.

   How many litres are needed to give everyone a milkshake?

   _____

3. The restaurant sells chocolate cake.

   How many cakes do you need to buy so that everyone has $\frac{1}{8}$ of a cake each?

   _____

   Is there any cake left over?

   _____

   If so, how much

   _____

4. Make up two more questions of your own involving fractions or decimals.

6 Decimals and fractions

1. How many ways can you describe, draw or use $\frac{3}{4}$ in a number sentence?

2. Two stars!

   Look through the pages in this Unit.

   Write two things that you are proud of learning in this Unit:

   * _____

   * _____

   And one wish!

   Write something in this Unit that you need to do more work on:

   * _____

7 Measurement, Area and Perimeter

Engage

What can we measure?

7A Estimating, measuring and recording length

Discover

1. Measure the height of each person in your group.

    • Record the heights in **centimetres (cm)**, and in **metres (m)** and **centimetres** in a table on a separate piece of paper.

2. You need: metre stick, calculator, tape measure, rough paper for working

    • Think of a way to measure the length of your step as accurately as possible.

    • Write the final measurement for each person in your group:

    _____

    _____

    _____

    _____

Step

3. Work with a partner.

    You need: your step length from question 2, calculator, trundle wheel, metre stick or tape measure

    • Choose a long distance to measure, for example: the length of the football pitch or the length of the school hall. Check your choice with your teacher.

    • Count the distance in steps:

    _____

    • Use a calculator to find the length in metres:

    _____

    • Measure the length again using the trundle wheel or metre stick:

    _____

7A Estimating, measuring and recording length

Explore

1. Draw arrows to match each of these lengths to the best **unit** of **measurement**: **kilometre (km)**, **metre (m)**, **centimetre (cm)** or **millimetre (mm)**. One has been done for you.

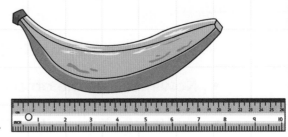

Length of an ant

Length of a football pitch

Height of a door

Length of a car

Distance across America

Length of a desk

Length of a book

Height of a tree

Length of a banana

Length of an orange pip

millimetres (mm)

centimetres (cm)

metres (m)

kilometres (km)

2. You need: metre stick

- Look in your classroom for objects between $\frac{1}{2}$ metre (50 cm) and 1 metre long.

- Choose an object.

- **Estimate** its length.

Measurement, Area and Perimeter

- Record your estimate in this table.

- Now measure the object to the nearest centimetre.

- Write the result in the table.

- Repeat with more objects.

- Check your partner's measurements.

Object	Estimated length	Actual length

3. Here are the heights of some jasmine plants in centimetres and in metres and centimetres.

- Complete this table to show the heights in two different ways.

Plant	Height (cm)	Height (m and cm)
1	115 cm	1 m 15 cm
2	234 cm	
3		1 m 75 cm
4	116 cm	
5	208 cm	
6	104 cm	
7		2 m 64 cm
8		2 m 3 cm

4. Who is correct?

- Circle the correct answer.

a) The height of a door

> I think it's about
> 4 metres.

> I think it's about
> 200 centimetres.

> I think it's about
> 100 centimetres.

b) The length of a pencil

> I think it's about
> 18 mm.

> I think it's about
> 20 m.

> I think it's about
> 18 cm.

c) The height of an elephant

> I think it's about
> 50 centimetres.

> I think it's about
> 3 metres.

> I think it's about
> 50 metres.

d) The length of your middle finger

> I think it's about
> 60 millimetres.

> I think it's
> 10 centimetres

> I think it's
> 6 millimetres.

5. Change the units. Remember 1 km = 1000 m.

a) 2 km = _____ m

b) 1.5 km = _____ m

c) _____ km = 7300 m

d) _____ km = 3400 m

e) 9.1 km = _____ m

f) 5.6 km = _____ m

7B Estimating, measuring and recording mass

Discover

1. Choose six similar toys.

   - Hold the toys in your hands to feel how heavy they are.

   - Try to estimate the order from lightest to heaviest.

   - Record your order in this table.

   - Estimate the **mass** in grams.

   - Now use the balance to find the exact mass of each item.

	Order	Estimate (grams)	Accurate mass (grams)
1			
2			
3			
4			
5			
6			

(left side label: Lightest to heaviest)

It's not easy! Continue to practise and you will improve.

2. Which is heavier: a small container full of sand or an identical container full of water?

   - Write a prediction: 'We think that ... because ...'

   We think that:

   _____

   _____

   - Discuss with your partner how you can test your prediction. Then try it out.

   What did you find out?

   _____

   _____

3. Try these problems. Show how you worked them out.

a) A coin has a mass of 20 g.

What is the mass of 8 coins?

b) A crate of apples has a mass of 5 kg.

The empty crate is 400 g.

What is the actual mass of the apples?

c) A cake has a mass of 900 g.

You cut it into six pieces.

What is the mass of each piece?

7B Estimating, measuring and recording mass

Explore

1. This table shows the masses of different fruits:

Lemon	Orange	Apple	Pineapple	Banana	Melon	Lime	Mango
110 g	300 g	160 g	890 g	170 g	1000 g	90 g	260 g

a) Which fruits have a mass of less than $\frac{1}{4}$ kg?

_____

b) Which fruits have a mass greater than $\frac{1}{2}$ kg?

_____

c) Which fruits have a mass between $\frac{1}{4}$ kg and $\frac{1}{2}$ kg?

_____

d) Which two fruits have a total mass of 1 kg?

_____

e) Which two fruits have the same mass as the mango?

_____

f) Which three fruits have a total mass of $\frac{1}{2}$ kg?

_____

2. Complete the scales in both grams and kilograms.

0 kg   0.4 kg   1.0 kg

0 g   200 g   600 g

3. You need: a set of metric masses, an electronic balance

- Collect six items from the classroom that you think have a mass less than a kilogram.

- Hold a 1kg mass in one hand and an object in your other hand. Estimate the mass of the object in grams.

- Record your estimate.

- Place the object on the electronic balance.

  Record the exact mass in the table.

- Repeat with the other items.

Object	Estimate (grams)	Actual mass (grams)

Did your ability to estimate the mass of the items improve?

_____

_____

7C Estimating, measuring and recording capacity

Discover

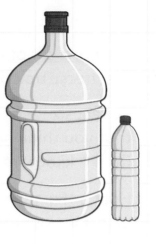

1. You need: a 5 litre container, a 1 litre bottle

   - Imagine that you pour a **litre** of water into the large container. Where will the water reach?

   - Place a sticky note to show your estimate.

   - Check your estimate by pouring in 1 litre of water.

   - Move the sticky note to the correct level.

   - Estimate the level for another litre and repeat the process.

   - Now add estimates for the remaining 3 litres.

   - Add the water and correct the placing of the sticky note.

   Are you getting better at estimating? It takes practice!

2. You need: measuring cylinder, water, stone, string or wire to hold the stone

   - Fill the **measuring cylinder** to about **half full**.

   - Record the water level in the table.

   - Lower the stone into the measuring cylinder. You will see that the water level rises.

Level at the start	
New level when stone added	
Difference	

   - Record the new water level.

   - Work out the difference.

   The increase in the level is equal to the amount of space the stone takes up.

   Archimedes discovered this more than 2000 years ago!

Explore

1. For each pair, circle the larger measurement.
   Two are done for you.

$\frac{1}{2}$ litre	(550 ml)	350 ml	$\frac{1}{4}$ litre	$\frac{3}{4}$ litre	700 ml	450 ml	($\frac{1}{2}$ litre)
I litre	850 ml	$\frac{1}{4}$ litre	140 ml	I litre	1100 ml	$\frac{3}{4}$ litre	850 ml

2. How much do you need to add to each of these containers to fill
   them to I litre?

   The first one has been completed for you.

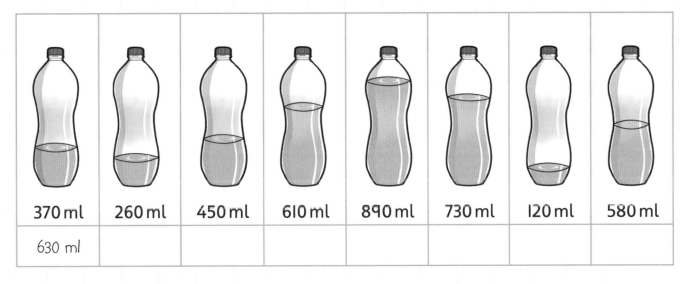

370 ml	260 ml	450 ml	610 ml	890 ml	730 ml	120 ml	580 ml
630 ml							

3. Complete the table.
   Show the **capacity** of these liquids in **millilitres** and litres.

	Capacity (millilitres)	Capacity (litres)
Apple juice		0.4 litre
Olive oil	800 ml	
Lemonade	1100 ml	
Orange juice		1.2 litres
Carton of milk	2000 ml	
Bottle of water		3.5 litres

Measurement, Area and Perimeter

147

7D Using and reading scales

Discover

1. This is a spring balance or newton meter.

   You use them in science lessons.

   They measure force in newtons or mass in grams.

   - Use the grams scale.

   - Look carefully at the scale to see what each interval is.

   - Be careful not to stretch the spring or you may damage it.

   - Estimate the mass of a pencil case.

   - Now use the newton meter to find the mass.

   - Record the result.

   - Repeat with other pencil cases.

Whose pencil case?	Estimate (grams)	Exact mass (grams)

2. We measure very small lengths in millimetres.

   Remember: 10 mm = 1 cm

   - Use your ruler to measure some small objects.
   - Write their lengths in the table.
   - Choose two of your own objects to add to the table.

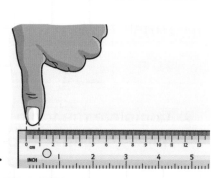

Object	Length in millimetres	Length in centimetres
Width of a pencil	6 mm	0.6 cm
Width of thumbnail		
Thickness of your textbook		

3. It is quite difficult to measure **to the nearest millimetre**.

   Measure these lines as accurately as you can:

   a) ___   Length: _____

   b) ____   Length: _____

   c) _   Length: _____

   - Try to draw these lines accurately. Use a sharp pencil.

   d) 12 mm

   e) 7 mm

   f) 5 mm

4. You need: large container of red water, large container of yellow water,
   100 ml measuring cylinder, 4 large test tubes, test tube rack

   Make 100 ml of different red/yellow mixtures:

	Amounts of red/yellow water	How much red do you need to make 100 ml total?	How much yellow do you need to make 100 ml total?	Colour of mixture
Mixture 1	1 part red to 1 part yellow			
Mixture 2	2 parts red to 3 parts yellow			
Mixture 3	3 parts red to 2 parts yellow	60 ml	40 ml	
Mixture 4	1 part red to 3 parts yellow			

What do you notice about the mixtures?

_____

_____

Make sure you tidy up neatly and wipe up any spills.

Measurement, Area and Perimeter

7D Using and reading scales

Explore

1. For each scale, write the difference between the two arrows.

   Suggest what you can use the scale to measure.

For example:

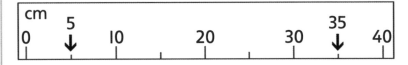

Difference: _____30 cm_____

I can use this scale to measure: _____the length of a line_____

a)
| kilogram |
| 0 ↓ 2   4   6   8   10 ↓ 12   14   16 |

Difference: _____

I can use this scale to measure: _____

b)
| metre |
| 0   ↓   50   ↓   100 |

Difference: _____

I can use this scale to measure: _____

c)
| millilitre |
| 0   ↓   500   ↓   1000 |

Difference: _____

I can use this scale to measure: _____

d)
| gram |
| 0 ↓   200   400 ↓ 600   800 |

Difference: _____

I can use this scale to measure: _____

e)

Difference: _____

I can use this scale to measure: _____

2. Draw a line to match the parcel to the correct scales:

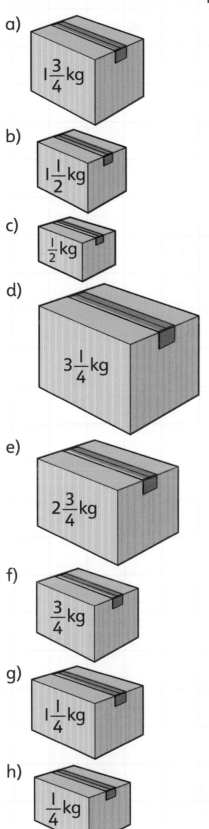

a) $1\frac{3}{4}$ kg

b) $1\frac{1}{2}$ kg

c) $\frac{1}{2}$ kg

d) $3\frac{1}{4}$ kg

e) $2\frac{3}{4}$ kg

f) $\frac{3}{4}$ kg

g) $1\frac{1}{4}$ kg

h) $\frac{1}{4}$ kg

3. The metric weights are

1 kg, 500 g, 2 × 200 g, 100 g in iron and 50 g, 2 × 20 g, 10 g and 5 g in brass.

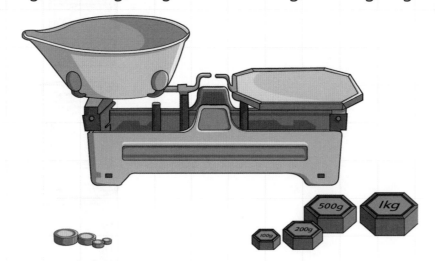

- Use as few weights as possible to balance these.
  The first one has been done for you.

a) 420 g: _____ 200 g, 200g, 20g _____

b) 1 $\frac{1}{4}$ kg: _____

c) 810 g: _____

d) $\frac{3}{4}$ kg: _____

e) 145 g: _____

f) 1605 g: _____

7E Drawing rectangles and calculating perimeters

Discover

- Use squared paper to carry out these investigations.

1. Draw as many different rectangles with a **perimeter** of 20 cm as you can.

   Work out the number of squares (the **area**) for each one.

   - Write the areas here: _____

   How many did you find? _____

   Do you think that you have found them all? _____

   - Check your findings with another pair of students.

2. Draw as many different rectangles with an **area** of 24 squares as you can.

   Work out the **perimeter** of each one.

   - Write the perimeters here: _____

   How many did you find? _____

   Do you think that you have found them all? _____

   - Check your findings with another pair of students.

3. Measure the perimeter of three rectangular objects in your classroom.
   For example: your book or your desk.

Object	Length	Width	Perimeter

7E Drawing rectangles and calculating perimeters

1. Measure the **edges** of these rectangles and calculate the perimeter:

a)

The perimeter is

_____ cm.

b)

The perimeter is

_____ cm.

c)

The perimeter is

_____ cm.

d)

The perimeter is

_____ cm.

2. Floor tiles cost $25 per **square metre**.

How much does it cost to tile these rooms?

- Draw a floor plan for each one.

a) Room 1 is 5 m long and 3 m wide.

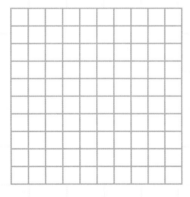

Cost: _____

b) Room 2 is 4 m square.

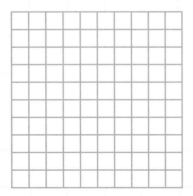

Cost: _____

c) Room 3 is 4 m long and $2\frac{1}{2}$ m wide.

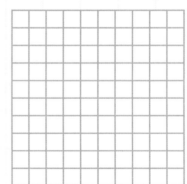

Cost: _____

3. Look back at question 2. Imagine that these are floor rugs instead of tiles, and the shopkeeper wants to give them a ribbon edge. Ribbon costs $6 a metre.

How much does it cost for ribbon to be added around the edge of each rug?

a) Room 1 rug:

b) Room 2 rug:

c) Room 3 rug:

7F Finding areas of rectangles

I. Estimate the length and width of your classroom.

- Calculate the estimated perimeter and area of your classroom.

Measure the length and width to the nearest metre.

- Calculate the perimeter and area of your classroom.

- Complete the table.

	Estimate	Measurement
Length		
Width		
Perimeter (m)		
Area (m²)		

2. Choose one of these areas:

| $30\,cm^2$ | $36\,cm^2$ | $40\,cm^2$ | $48\,cm^2$ |

- Try to predict the **measurements** of the rectangle or square that has the smallest perimeter.

- Try to predict the measurements of the rectangle or square that has the largest perimeter.

- Write your predictions.

  I choose to investigate this area: _____ cm²

  I think the rectangle with the smallest perimeter will be _____

  cm long and _____ cm wide.

  I think the rectangle with the largest perimeter will be_____ cm

  long and _____ cm wide.

- Now test your predictions by drawing all the possible rectangles on squared paper and calculating the perimeters.

7F Finding areas of rectangles

Explore

1. How many **square centimetres** are in a **square metre**? _____

2. Find the area of each of these shapes in squares.

   Then draw a different shape with the same area.

   My shape

a)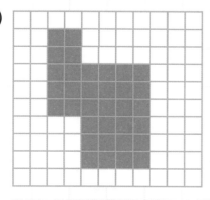

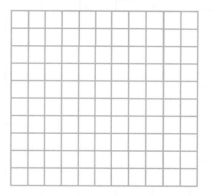

The area is _____ squares.

b)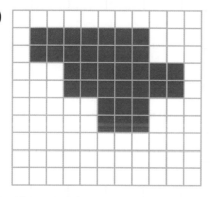

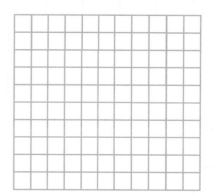

The area is _____ squares.

c)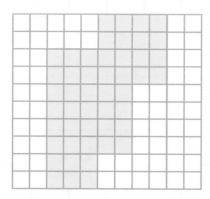

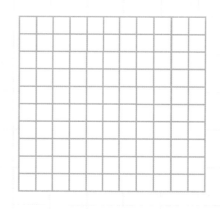

The area is _____ squares.

d)

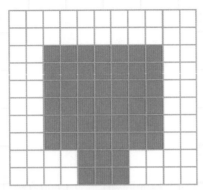

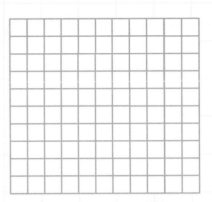

The area is _____ squares.

e)

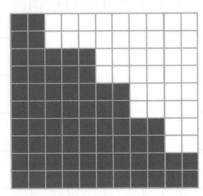

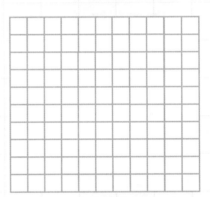

The area is _____ squares.

3. Complete this table. You can draw the shapes on squared paper.

Length × width	Perimeter	Area
1cm × 1cm	4 cm	1 cm²
1cm × 2cm		
1cm × 3cm		
1cm ×		

158

Can you see any patterns in the answers? Write what you notice:

_____

_____

_____

Connect

Plan a pizza supper!

Cheese and Tomato Pizza Recipe

Ingredients for 1

75 g flour

10 g butter

30 ml milk

4 cherry tomatoes

25 g cheese

salt and pepper

herbs

Use poster paper to write out your new recipes.

Plan a pizza supper for 10.

- Use the recipe to calculate how much of each ingredient you need for 10 people.

- Decide what drink you would like. How much do you need?

Plan a pizza supper for 100.

- Calculate how much of each ingredient you need for 100 people.

- Change the amounts to kilograms.

- How much of your chosen drink do you need for 100 people?

7 Measurement, area and perimeter

1. Write the meaning of these prefixes:

    a) kilo- _____

    b) centi- _____

    c) milli- _____

2. Choose the correct unit to measure:

    a) The length of the classroom _____

    b) The distance to the moon _____

    c) The mass of a cent _____

    d) The mass of a human _____

    e) The capacity of a cup _____

    f) The capacity of the petrol tank in a car _____

3. Draw three things that have a measuring scale.

    Label each one. Explain what and how they measure:

8 Time

Why do we measure time?

How do we measure time?

8A Telling the time

Discover

I. Think about your day. Fill in the times:

	Time in words	Digital	Analogue
I get up at:			
I have breakfast at:			
I arrive at school at:			
Lessons start at:			
Playtime is at:			
School finishes at:			
I arrive home at:			
I eat at:			
I go to bed at:			

2. Continue these time sequences.

> **For example:**
> - 3.45 a.m., 4.00 a.m., 4.15 a.m., <u>4.30 a.m.</u>, 4:45 a.m., <u>5.00 a.m.</u>

- 4.00 a.m., 4.20 a.m., 4.40 a.m., _____, _____, _____

- 7.10 p.m., 9.10 p.m., 11.10 p.m., _____, _____, _____

- 9.40 a.m., 10.40 a.m., 11.40 a.m., _____, _____, _____

- 10.35 a.m., 11.05 a.m., 11.35 a.m., _____, _____, _____

- 5.20 p.m., 5.30 p.m., 5.40 p.m., _____, _____, _____

3. Times around the world. It is 4.00 p.m. in London. What time is it in:

- Anchorage in Alaska (9 hours behind London) _____

- Guadalajara in Mexico (6 hours behind London) _____

- Sao Paulo in Brazil (3 hours behind London) _____

- Paris in France (I hour ahead of London) _____

- Doha in Qatar (3 hours ahead of London) _____

- Bangalore in India ($5\frac{1}{2}$ hours ahead of London) _____

- Osaka in Japan (9 hours ahead of London) _____

- Wellington in New Zealand (12 hours ahead of London)
  _____

Did you remember to add a.m. or p.m.?

8A Telling the time

Explore

1. These cards show the events in a footballer's day.

   • Work with a partner.

   • Put the events in the correct order.

H

Drive home at 7.30 p.m.

G

Second half of match at 6.00 p.m.

L

Pre-match warm-up at 4.50 p.m.

K

Post-match discussion at 6.56 p.m.

E

Team talk by the manager at 4.35 p.m.

I

Training session at 10.05 a.m.

J

Rest and physiotherapy at 2.35 p.m.

B

Extra time at 6.45 p.m.

C

Drive to football stadium at 9.00 a.m.

A

Half-time oranges at 5.45 p.m.

F

First half of match at 5.00 p.m.

D

Lunch at 1.15 p.m.

Time order	Card letter
1st card	C
2nd card	
3rd card	
4th card	
5th card	
6th card	
7th card	
8th card	
9th card	
10th card	
11th card	
12th card	

2. Here are some clocks showing times between 5 o'clock and 6 o'clock.

- Complete this table.

- Write the number of **minutes** *past* 5 o'clock and the number of **minutes** *to* 6 o'clock.

The first two are done for you as examples.

	Number of minutes *past* 5 o'clock	Number of minutes *to* 6 o'clock	How do you say this time in words?
	8	52	8 minutes past 5
	56	4	4 minutes to 6

3. What time is it? Fill in the times on the digital and analogue clocks.
The first row is done for you.

It's 3 minutes to 4.	3:57	(clock showing 3:57)
It's 20 minutes past 9.		
It's 27 minutes past 7.		
It's 25 minutes to 10.		
It's 9 minutes to 3.		
It's 14 minutes past 11.		
It's 6 minutes past 4.		
It's 12 minutes to 6.		

8B Timetables and calendars

Discover

I. Here is a calendar for 2050:

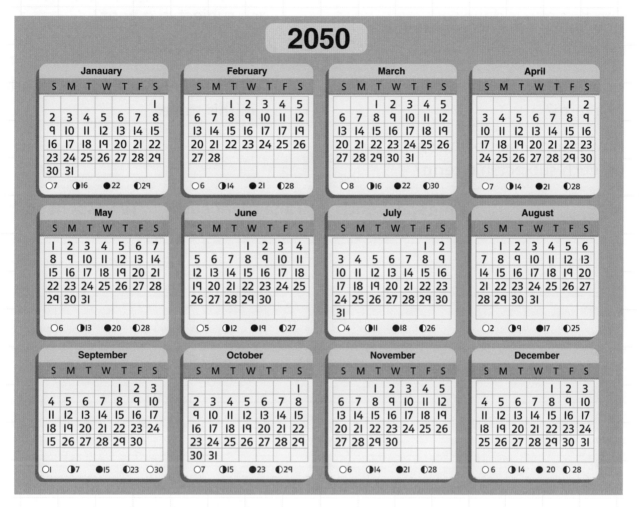

a) How old will you be in 2050?

_____

b) Is 2050 a **leap year**?
   How do you know?

_____

c) What day is 26 March?

_____

d) Which months have five
   Wednesdays?

_____

e) Which months start on
   a Tuesday?

_____

f) Which month ends on a Sunday?

_____

g) How many full moons are there
   in the year 2050?

_____

h) What day will 2051 start on?

_____

2. Look at your school class **timetable**.

How much time do you spend on each subject in a **week**?

How many weeks are there in a term?

How much time do you spend on each subject in a **term**?

How much time do you spend on each subject in a **year**?

- Use paper for your working.

- Then write your findings clearly in this table.

Subject	Time spent each week	Time spent each term	Time spent each year

3. Do some children at your school travel by school bus?

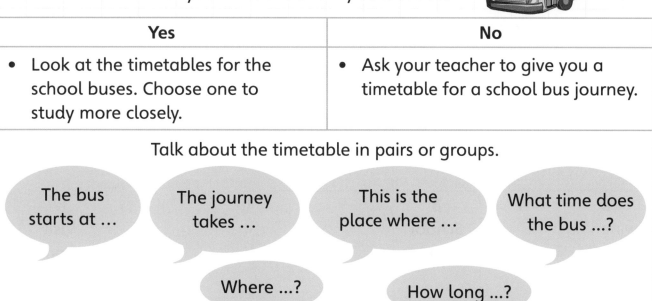

Yes	No
• Look at the timetables for the school buses. Choose one to study more closely.	• Ask your teacher to give you a timetable for a school bus journey.

Talk about the timetable in pairs or groups.

The bus starts at …

The journey takes …

This is the place where …

What time does the bus …?

Where …?

How long …?

• Write six questions about the timetable.

Q1. _____

_____

Q2. _____

_____

Q3. _____

_____

Q4. _____

_____

Q5. _____

_____

Q6. _____

_____

• Ask another pair your questions.

Time

169

8B Timetables and calendars

Work with a partner.

I. Here is a friend's calendar for the month of June.

Sunday	Monday	Tuesday	Wednesday	Thursday	Friday	Saturday
		1	2	3	4	5
6 Swimming Club	7	8 School trip	9	10	11	12 Visit grandparents
13 Swimming Club	14	15	16 School theatre visit	17	18	19
20 Swimming Club	21	22	23	24 End of term	25	26 Cinema with friends
27 Swimming Club	28	29	30			

What is the date of the last day of term? _____

When is your friend visiting the cinema? _____

What activity does your friend do every week? _____

What is happening on 12 June? _____

When is the school trip? _____

What special event is on a Wednesday? _____

2. A family goes to London on holiday. One day they visit Oxford by train.

Timetable:

Train	Depart London	Arrive Oxford	Journey time
A	8.22 a.m.	9.20 a.m.	58 minutes
B	8.51 a.m.	9.53 a.m.	
C	9.00 a.m.	10.04 a.m.	
D	9.21 a.m.	10.18 a.m.	

Train	Depart Oxford	Arrive London	Journey time
E	5.31 p.m.	6.28 p.m.	
F	5.43 p.m.	6.54 p.m.	
G	6.07 p.m.	7.16 p.m.	
H	6.31 p.m.	7.32 p.m.	

- Work out the journey times for each train. Write them in the timetable.

Which is the quickest train to Oxford? _____

Which is the quickest train back to London? _____

The family arrives at the station in London at half past 8.

What time is the next train? _____

How long does that train take to reach Oxford? _____

The family wants to be back in London by 7 o'clock.

Which trains can they choose? _____

They decide to take the quicker train.

What time will they get back to London? _____

How long did they spend in Oxford? _____

8C Measuring time intervals

Discover

1. Look at these different measurements for time.

    What can you measure using each **unit of time**?

    Here are two examples. Write more examples.

    - Years _____   _____*age of people*_____

    - Minutes _____   _____

    - Hours _____   _____

    - Months _____*school term*_____   _____

    - Seconds _____   _____

2. What is your **date of birth**?   What is your partner's date of birth?

    _____   _____

    Exactly how old are you in years?   Exactly how old is your partner
                                            in years?
    _____
                                            _____
    Exactly how old are you in months?
                                            Exactly how old is your partner
    _____   in months?

                                            _____

3. Just a minute! Work with a partner and use a stopwatch to time each other.

    - Close your eyes and keep them closed for what you think is a minute.

      How close to a minute was your time? _____

      How many times can you write your name in a minute? _____

      How many times can you touch your toes in a minute?

      Stand up straight after each touch! _____

    - Sit down. Stand up when you think a minute has passed.

      Is a minute a long or short time? _____

8C Measuring time intervals

1. Which unit of time do you use to measure:

    - The time to eat breakfast _____

    - The time to run a marathon (26 miles!)_____

    - The time for a baby to grow up _____

    - The time to write your first name _____

    - The time for a football match _____

    - The time to drive 500 km _____

    - A school year? _____

2. Here is a hospital doctor's appointment list for one day:

Tuesday's appointments	
10.00 a.m.	Mr J. Jupiter
10.45 a.m.	Mr S. Saturn
11.10 a.m.	Mr M. Mars
11.55 a.m.	Mr P. Pluto
1.45 p.m.	Miss N. Neptune
2.05 p.m.	Miss M. Mercury
2.35 p.m.	Miss U. Uranus
3.05 p.m.	Miss V. Venus

    How long was Mr Jupiter's appointment? _____

    Mr Pluto was 20 minutes late. What time did he arrive? _____

    Miss Venus' appointment lasts 70 minutes.

    What time does her appointment end? _____

    Which patient has the shortest appointment? _____

    Mr Mars arrived 20 minutes early. What time did he arrive? _____

Miss Neptune arrived at 1. 34 p.m. Was she early or late for

her appointment? _____

By how long? _____

Mr Saturn arrived exactly on time. The journey took 55 minutes.

What time did he leave home? _____

Miss Mercury left home at 1.25 p.m. and took 45 minutes to reach the

hospital. What time did she arrive at the hospital? _____

How late was she for her appointment? _____

3. • Change to minutes:

    a) 120 seconds _____

    b) $3\frac{1}{2}$ hours _____

    c) 360 seconds _____

    Change to days:

    d) 3 weeks _____

    e) 72 hours _____

    f) 9 weeks _____

• Change to weeks:

    g) $\frac{1}{2}$ year _____

    h) 49 days _____

    i) 2 years _____

• Change to years:

    j) 36 months _____

    k) a century _____

    l) 5 decades _____

8 Time

- Plan a whole day's schedule for a children's TV channel.

- Include a variety of programmes, for example: news, cartoons, adventure, wildlife …

For each programme, show:

- its title

- its start and finish time

- its length in minutes.

- Use suitable paper for your planning and your final design.

8 Time

1. Why are these numbers important in learning about time?

   ## 7   60   12   365   24

   I think 12 is important because _____

   60 is important because there are _____

   7 is the number of _____

   365 is the number of _____

   24 is the number of _____

   Discuss with a partner other numbers that are important in time.

2. Here are some times. Write each time in words. The first is done as an example.

   7.45 a.m. _____*a quarter to eight*_____

   9.15 a.m. _____

   12.30 p.m. _____

   3.45 p.m. _____

   7.40 p.m. _____

3. What happens in your school at these times?

   7.45 a.m. _____

   9.15 a.m. _____

   12.30 p.m. _____

   3.45 p.m. _____

   6.00 p.m. _____

4. Think about why timetables are useful.

How does shape make our world a more interesting place to live?

9A 2D shapes and classifying polygons

Discover

I. Use six identical **regular triangles (equilateral triangles).**

- Put them together to make a single **polygon.**

  There are lots of different ways to do this.

- Use a whole number of triangles.

- Draw the shapes listed on this grid:

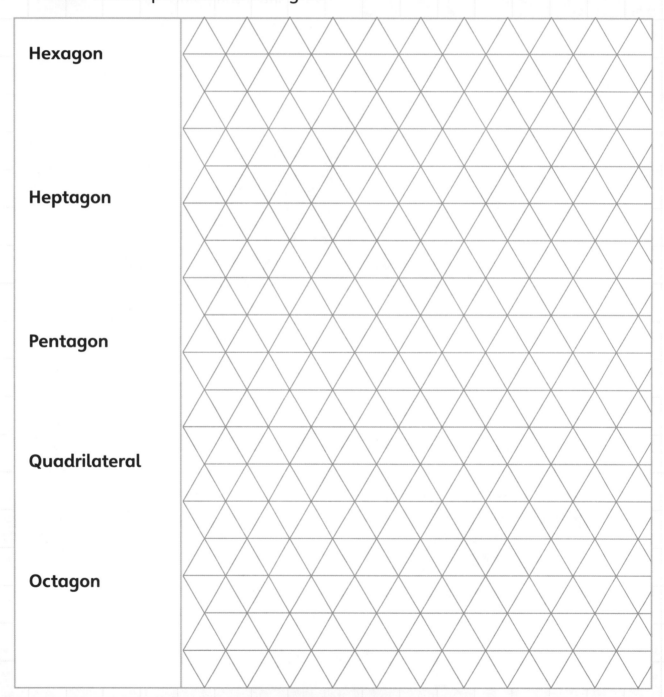

Hexagon

Heptagon

Pentagon

Quadrilateral

Octagon

2. Look for different polygons in this grid.

 • Use different coloured pencils to show these shapes:

   • a pentagon

   • a triangle

   • a hexagon

   • a quadrilateral

   • a heptagon.

9A 2D shapes and classifying polygons

My partner today is _____.

I. Use a ruler to draw four different-shaped triangles.

Try to make them look as different as possible.

- Look at your partner's triangles.

Do any of them look like yours?

_____

How can you describe a triangle so that someone can draw another one that is exactly the same?

Talk about this question with your partner.

- Write your ideas here:

_____

_____

_____

Mathematicians use special words to classify different kinds of triangles.

- Fill in the definitions.

Equilateral triangles have _____

Isosceles triangles have _____

Scalene triangles have _____

Look back at the triangles you drew at the beginning of the lesson.

With your partner, decide which kind of triangle each one is: equilateral (E), isosceles (I) or scalene (S).

- Write E, I or S inside each triangle.

- Draw one of each type of triangle. Label them E, I and S.

2. Look carefully at these shapes:

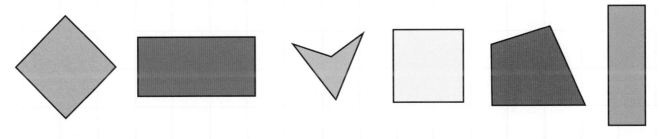

- These shapes are all _____ because they all

  have _____ sides.

- Write an R in the shapes that are regular.

- Write an O in any shapes that are **oblongs**.

- Draw a circle around any **concave** shapes.

9B 3D shapes

1. With your partner choose 10 different **3D (three-dimensional) shapes** for sorting.

   Decide a way to sort and classify them using a Venn diagram.

   • Draw and label the circles of the Venn diagram below.

   • Place the shapes in the correct part of the diagram.

   Are there any shapes in the intersection (where the circles overlap)?

   What are they?

   _____

Ask your teacher to take a photograph of your work to stick into your book.

9B 3D shapes

I. Look carefully at the 3D shapes.

- Complete this table.

  You can use the shapes to help you.

	Name of shape	Number of faces	Number of vertices	Number of edges

2. Describe these shapes:

It has one flat face and one curved one. It has one vertex. It is a cone.				

3. Work out which solid shape matches each clue.

- Find another shape that also matches. Use the solid shapes to help you.

What can it be?	It is a ...	or a ...
This shape has triangular and rectangular faces		
This shape has six vertices		
This shape has four triangular faces		
This shape has a curved face		
This shape has more than four rectangular faces		

9C Line symmetry

Discover

I. Use these grids to design four **symmetrical** cartoon creatures for a new computer game.

- Use at least three colours to design each creature.

- Make each creature quite different.

- Draw in the **line of symmetry** for each one.

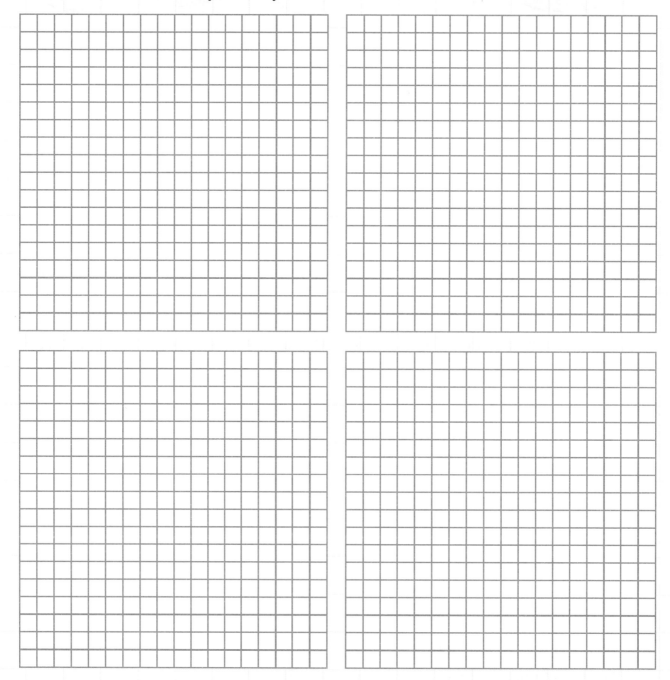

2. Symmetrical buildings are pleasing to look at.

Architects have been designing symmetrical buildings for hundreds of years.

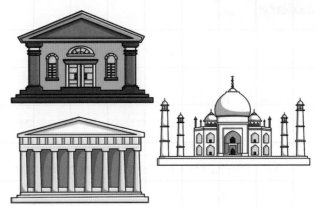

- Use this grid to design the front of an interesting symmetrical building:

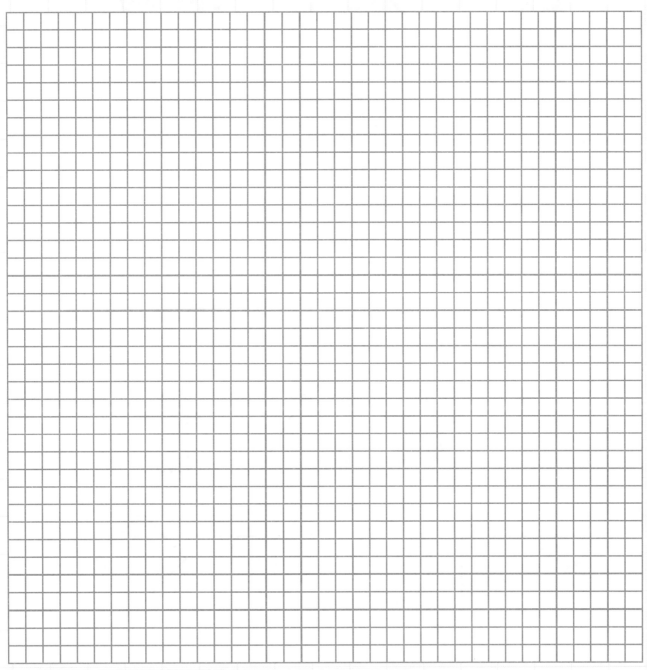

9C Line symmetry

I. Read this sentence carefully:

> The number of lines of symmetry in a regular polygon is equal to the number of sides of the polygon.

- Draw all the lines of symmetry on these polygons to test this statement:

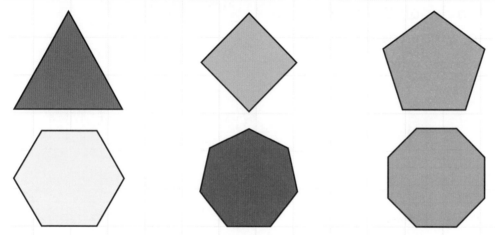

Do you agree with the statement? _____

2. Look at the two lines on each shape.

One is a line of symmetry. The other is not.

- Put a tick ✓ beside the line that you think is a line of symmetry.

  Are you correct? Use a mirror to check.

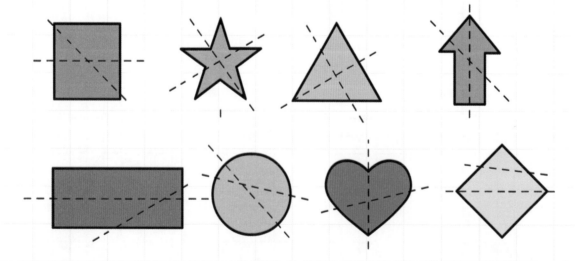

3. Write letter names sorting these shapes into the correct part of the Carroll diagram:

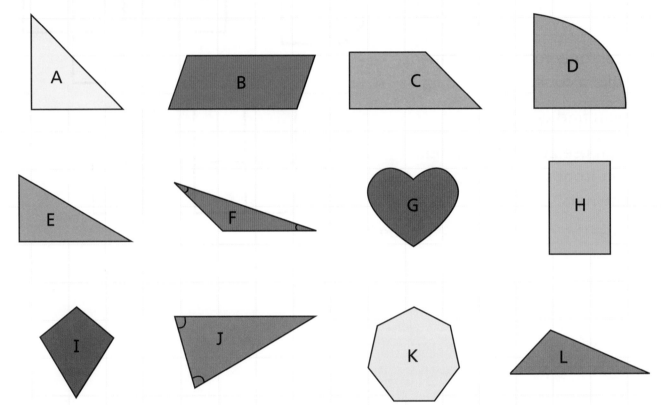

	Right angle	No right angle
At least one line of symmetry		F
No line of symmetry		

Discover

A cube has six square faces.

You can arrange the faces in many different ways.

You can only fold some of them to make a cube.

Can you find which ones?

You can use 2D shapes or make models from squared paper to test out the **nets**.

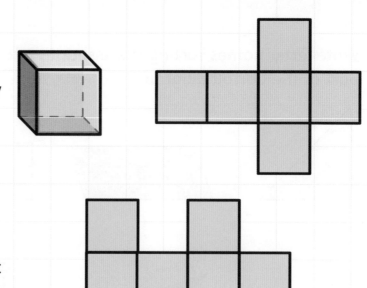

• Draw the nets that make a cube here:

9D 2D nets of 3D shapes

Name the shape made by each of these nets:

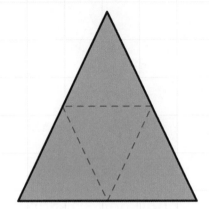

a) _____

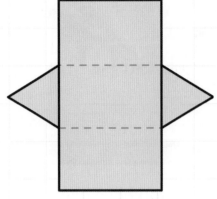

b) _____

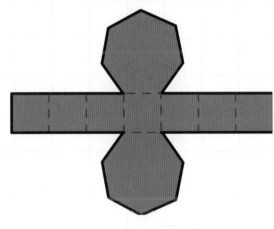

c) _____

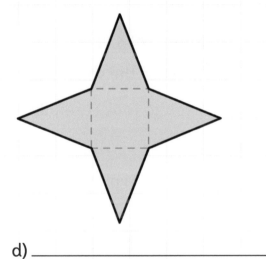

d) _____

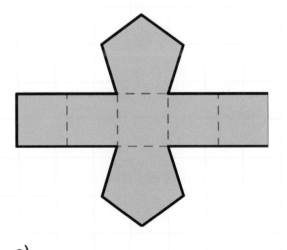

e) _____

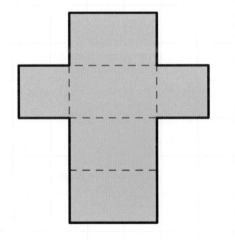

f) _____

9 Shape and geometry

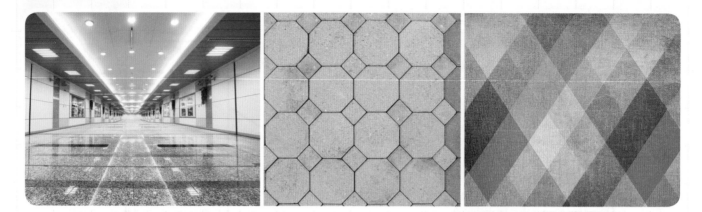

In this unit, you have been looking at **2D (two-dimensional)** and 3D shapes and symmetry.

- Design a maths trail around your school that includes these things.

In your group, walk around the school.

- Make a list of the shapes and examples of symmetry that you see.

You will need to look carefully – up and down – and use your imagination.

For example:

- Light fittings may be interesting symmetrical shapes.
- Floor tiles may have regular polygons.
- Wallpaper or curtains may have symmetrical patterns.

- Decide on a route to include the best things you have seen.

Try to find some unusual shapes.

You may be able to add some items to the route so that your trail has more variety. For example: put a collection of different **pyramids** on a bookcase and ask a question about them.

- Write instructions for your route for other groups of students to follow.

Aim for 10 good maths questions. You need to be clear about:

- **where** to be

- what to **look** at

- **what** to find out.

Here are some example questions:

Start at the Main Entrance. Look at the glass vase on the table.

What shape is it?

Draw the net of the vase.

Go to the Dining Room.

Look at the student tables.

What shape are they?

Find a piece of cutlery with line symmetry.

Sketch it here.

Draw the line of symmetry.

Go to the Adventure Playground.

Look at the equipment and look for prisms.

Sketch and label what you see.

- Swap your completed trail with another group.

- Try out each other's trails. Enjoy!

9 Shape and geometry

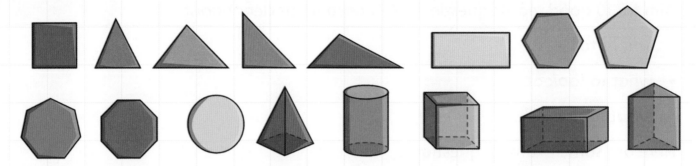

Starting with the terms '2D shapes' and '3D shapes', can you use words and diagrams to link all the work covered in this Unit?

There are some useful diagrams and words on this page.

2D shapes

3D shapes

prism   polygon   tetrahedron   faces   net   edges   oblong   heptagon
pyramid   regular   isosceles triangle

10 Position and Movement

What is an **angle?**

Which of these angles are **right angles?**

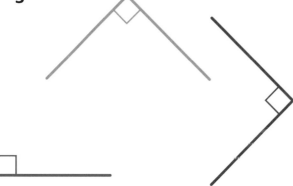

I think the one with red lines is a right angle.

I think they are all right angles.

How can we check?

Are these angles the same size?

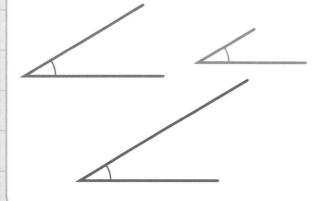

Yes, they are the same.

No, they are not.

10A Measuring angles

Discover

Open the geo-strips to make an angle.

Carefully draw the angle on a piece of paper.

Move the geo-strips to make a different angle.

Draw it on paper.

Draw eight different sized angles in total.

Cut out your angles carefully.

Arrange them in order of size.

Where does a right angle fit in your order?

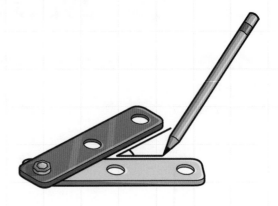

- Stick your angles here in order of increasing size:

10A Measuring angles

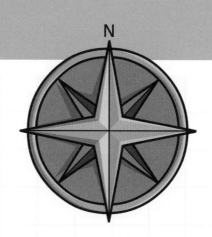

Explore

I. What direction are you travelling after the turn?

All turns are **clockwise**.
The first answer is done for you.

Travelling **north (N)**, turn 90°. You are now travelling: _____East_____	Travelling **north**, turn 180°. You are now travelling: _____	Travelling **west (W)**, turn 90°. You are now travelling: _____	Travelling **south-east (SE)**, turn 180°. You are now travelling: _____
Travelling **north-east (NE)**, turn 90°. You are now travelling: _____	Travelling **north-west (NW)**, turn 180°. You are now travelling: _____	Travelling **south-east**, turn 90°. You are now travelling: _____	Travelling **east (E)**, turn 135°. You are now travelling: _____
Travelling **south-west (SW)**, turn 360°. You are now travelling: _____	Travelling **south (S)**, turn 135°. You are now travelling: _____	Travelling **south-east**, turn 270°. You are now travelling: _____	Travelling **north-east**, turn 135°. You are now travelling: _____

2. Write the correct direction letters on the points of this **compass**.

North is shown for you.

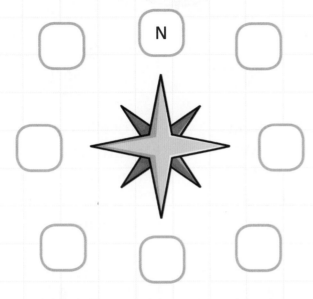

3. Use two throws of a dice to decide your starting direction and the angle you need to turn.

All turns are clockwise.
Complete the table.

First throw decides starting direction	
1	NE
2	SE
3	NW
4	SW
5	E
6	W

Second throw decides angle to turn	
1	45°
2	90°
3	135°
4	180°
5	225°
6	270°

Starting direction	Angle	Final direction
E	135°	SW

10B Giving directions to follow a path

Discover

My partner today is _____.

Work on your own to start with.

Then check that you agree with your partner's answers.

I. Follow the instructions. Start at the red cross.

Draw crosses on the grid to show your path.

- Forward 5 squares

- Turn 90° clockwise

- Forward 5 squares

- Turn 90° clockwise

- Forward 5 squares

- Turn 90° clockwise

- Forward 5 squares

What shape does the path make?

_____

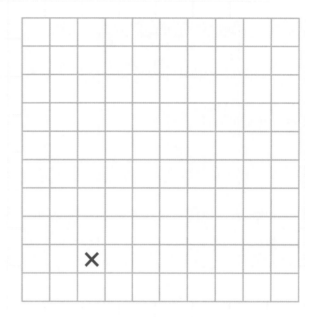

2. Write instructions to follow this path, starting at the red cross:

_____

_____

_____

_____

_____

_____

_____

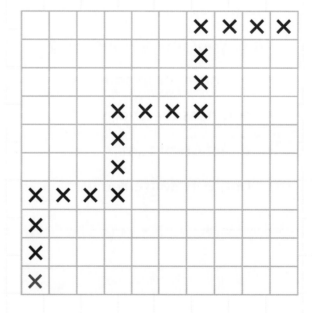

10B Giving directions to follow a path

Explore

Here is a shop floor plan.

White areas are walkways.

Blue areas are furniture, for example: display cabinets, tills.

- Work out two different routes to move from the Entrance to the Exit.

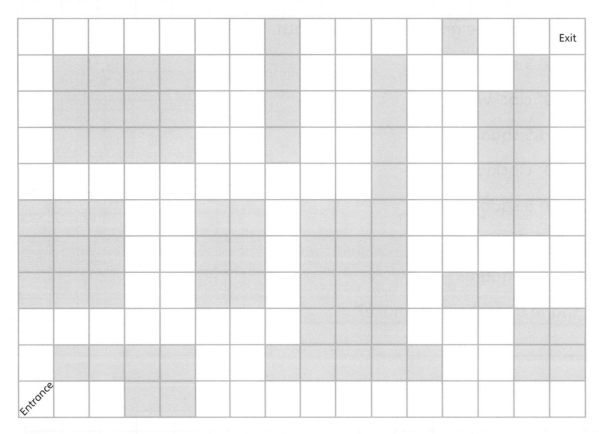

Write your instructions using: _____

- the number of squares   _____

- compass directions   _____

- direction of turn   _____

- angle in **degrees**.   _____

_____   _____

_____   _____

10C Coordinates of a square on a grid

Discover

I. **Find the Treasure Game!**

- Work with a partner.

- You each have:

  - five treasure chests (T)

  - three lots of gold bars (G)

  - two pirates (P).

- Choose where to put your treasure, gold and pirates on your map grid.

- Write T, G or P in the squares you choose. Don't show your partner your map.

- Take turns to choose a square on your partner's map, using coordinates.

Did you choose a square with treasure or gold? You get the treasure.

Did you choose a square with a pirate? You miss a turn!

The winner is the first to collect all their partner's treasure and gold.

8										
7										
6										
5										
4										
3										
2										
I										
	A	B	C	D	E	F	G	H	I	J

2. Look at this map:

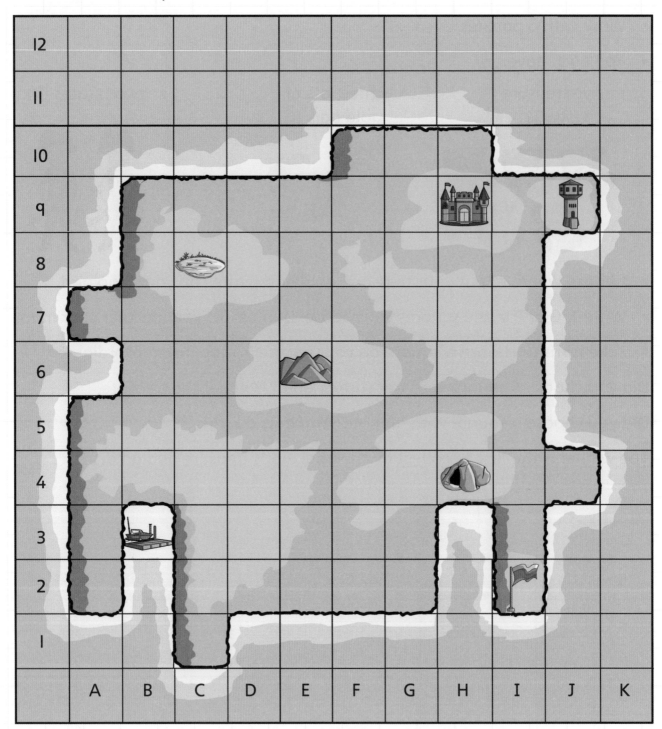

a) Write the coordinates of the:

- Cave   (_____ , _____)

- Harbour   (_____ , _____)

- Flag post   (_____ , _____)

- Hill   (_____ , _____)

- Castle   (_____ , _____)

- Lookout   (_____ , _____)

- Pond   (_____ , _____)

b) Draw on the map:

- two more hills at (C, 5) and (E,7)

- two more flag posts at (C,I) and (A,7)

c) Add two more features of your own to the map.

Write the coordinates for them:

Picture for feature	Description of feature	Coordinates

Often both the first and second coordinates are numbers.

The first number shows the position on the horizontal **axis**.

The second number shows the position on the vertical axis.

I. Look at this grid:

10								1.5		
9			42							
8					16					
7								2400		
6		$\frac{1}{2}$								
5				72						
4							63			
3								0.75		
2	400				$\frac{2}{3}$					
1										15
	1	2	3	4	5	6	7	8	9	10

- Work out the answer to each of these calculations.

- Match the answer to the correct coordinates for its place on the grid.

- Complete the table. The first is done for you as an example.

- Write two more calculations of your own. Try to make your questions difficult!

Calculation	Answer	Coordinates
9 × 7?	63	(7, 4)
$\frac{3}{4}$ as a decimal?		
45 ÷ 3?		
Half of 3?		
300 × 8?		
Double 36?		
The next number in this series: 4, 2, 1,...		
$1 - \frac{1}{3}$?		
2 × 2 × 2 × 2?		
7 × 6?		
197 + 203?		

2. Play this game with a partner.

- Use an 8-sided dice.

The first throw gives the first coordinate.

The second throw gives the second coordinate. No cheating!

Each pair of coordinates give you a score on the board.

- Take turns. Add up your scores.

The first person to reach 1000 is the winner.

8	100	0	10	0	0	20	200	30
7	120	400	0	0	0	0	70	350
6	0	60	0	10	350	0	0	0
5	0	0	0	0	200	0	450	0
4	0	200	20	0	0	100	40	80
3	50	0	0	160	0	500	0	250
2	300	150	0	0	0	0	110	0
1	0	0	140	0	20	0	0	90
	1	2	3	4	5	6	7	8

10 Position and movement

What angles can you see in this painting by the artist Wassily Kandinsky?

_____

_____

- Design your own angle picture on a separate piece of paper. Include:

  - three or more right angles

  - three or more straight lines

  - three or more angles smaller than 90°

  - three or more angles greater than 90°

  - three or more circles or parts of circles. Draw them using **compasses**.

After you draw your angles and shapes, use colour, pattern or different shapes to add interest.

10 Position and movement

You have lost your angle measurer and you want to draw some angles.

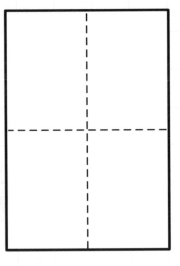

Using sheets of A4 paper, what angles can you make by folding?

Here are some to try:

- 90°

- 45°

- 30°

- 60°

What others can you make?

- Cut your angles out.

- Draw each angle here.

- Under each angles, write how you made it.

Handling Data

How can we collect, organise and present **data**?

What can data tell us?

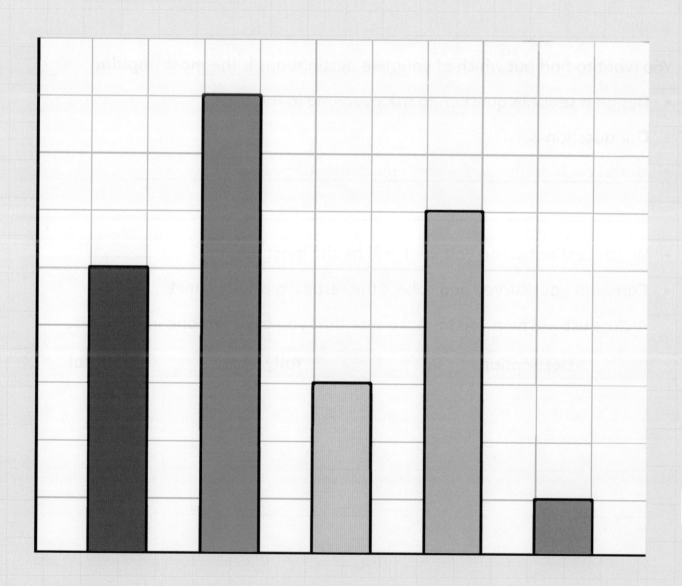

Discover

- Choose **one** of these topics to investigate:

  A. Where would you like to visit for a day?

  B. Where would you like to go for a school trip?

  C. What country in the world would you like to visit?

  Our group chose topic _____.

- Choose five popular destinations: _____

  _____

  _____

You want to find out which of your five destinations is the **most popular**.

- Design a sensible question to ask everyone in the class.

  Our question is:

  _____

  _____

- Which destination do you think will be the most popular? _____

- Carry out your **survey** and collect the results in a **tally chart**.

  A class **list** will help you to make sure that you ask everyone in the class.

Destination	Tally	Total

- Look at the data. Each student in your group chooses a different way to **represent** the data: a **bar chart**, a simple **table** or a **pictogram**.

- Draw your graph or table here:

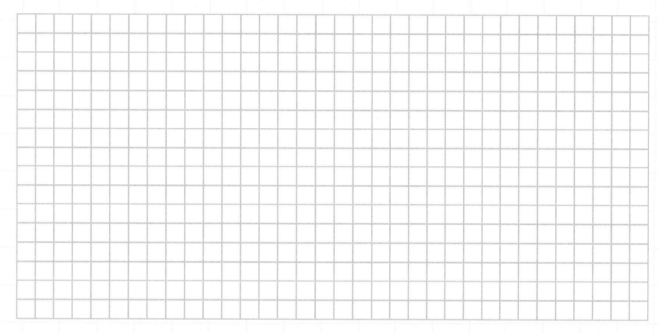

- Write a short report on the information your data provides.

  Include these points:

  - The most popular choice and the runner-up (second choice).

  - Was the result what you predicted?

  - Which form of chart or table shows the information most clearly?

  _____

  _____

  _____

  _____

  _____

  _____

11A Collecting, presenting and interpreting data

Explore

I. Here is a tally chart showing the sales of different flavours of ice-cream sold at an ice-cream shop.

a) Complete the 'Total' column.

Flavour	Tally	Total
Vanilla	卌 卌 卌 卌 卌 II	
Chocolate	卌 卌 卌 卌 III	
Strawberry	卌 卌 卌 I	
Mango	卌 II	
Coconut	卌 卌 卌 IIII	

b) Draw a bar chart for this data. **Label** the **axes** clearly and give your chart a **title**!

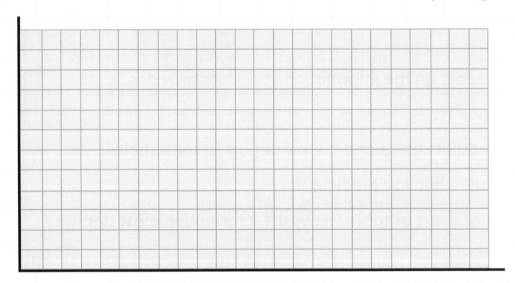

c) Write five facts from the bar chart:

_____

_____

_____

_____

_____

2. Students in Stage 4 had a holiday one day.

They collected information to see what everyone did on their day's holiday.

Where Stage 4 students went on their day's holiday

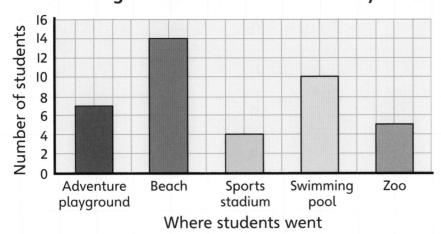

Tick ✓ the statements that you agree with:

a) Ten students chose to go to the swimming pool.

b) More students went to the adventure playground than the zoo.

c) Two more students went to the zoo than the sports stadium.

d) The beach was the most popular choice.

e) It is summer.

f) The sports stadium was a popular choice.

g) More than half the students went swimming.

h) There are 40 students in Stage 4.

Why do you disagree with the statements that you have not ticked?

_____

_____

i) Use the data to write two true statements of your own:

_____

_____

Discover

Here are some weather data from a country in Europe collected over three months:

Weather	Number of days
Sunny	35
Cloudy	21
Windy	9
Rainy	19
Stormy	7

- You are going to investigate the effect of different **scales** on bar charts.

- Agree with your group what **intervals** to draw.

- Make sure you each use **different** intervals.

- Give the bar chart a title.

- Label the axes and scales clearly.

Draw your bar chart on the next page:

- Compare the results in your group.

- Answer these questions:

Which scale do you think shows the data most effectively?

The best scale is the one with intervals of _____.

Do you all agree? _____.

Why do you think this?

_____

_____

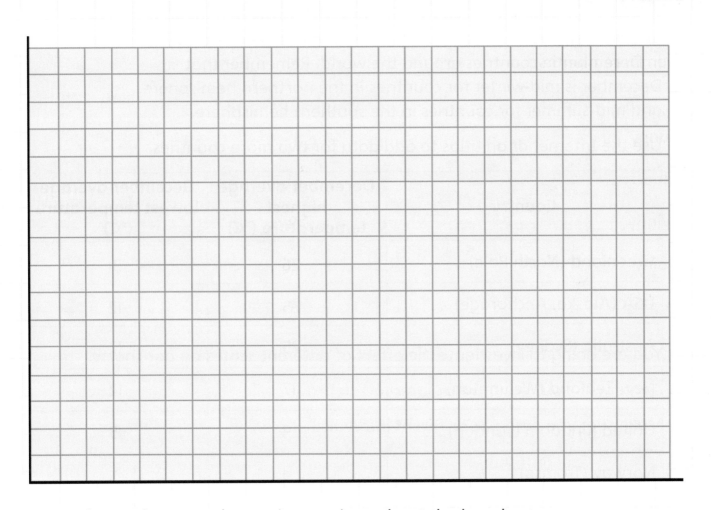

- Work together to make up six questions about the bar chart:

_____

_____

_____

_____

_____

_____

_____

11B Comparing scales with different intervals

I. This data shows the average hottest and coldest temperatures in December in countries around the world. Remember that December is mid-winter for countries in the northern hemisphere and mid-summer for countries in the southern hemisphere.

Use the Internet or an atlas to add data for two more countries.

Country	December average highest temperature (°C)	December average lowest temperature (°C)
Greenland (North Pole)	−26	−31
USA (Alaska, Anchorage)	−5	−12
Australia (Sydney)	25	17
New Zealand (Wellington)	17	12
United Kingdom (London)	9	5
Norway (Tromsø)	−1	−5
Malta (Valetta)	16	11
Iceland (Akureyri)	0	−4
Russia (Siberia)	−10	−17
Finland (Tampere)	−1	−6
Spain (Madrid)	10	4
Saudi Arabia (Jeddah)	29	20
United Arab Emirates (Dubai)	25	16
South Africa (Johannesburg)	24	15
Uganda (Entebbe)	26	16

Japan (Tokyo)	11	4
India (Cochin)	30	23
India (Shimla)	11	4
South Pole	−26	−28
Barbados	28	23
Brazil (Sao Paulo)	26	18

Which country has the highest temperature? _____

Which country has the lowest temperature? _____

What is the difference in temperature between highest and lowest

temperatures in Norway? _____

Which country has the biggest difference between the highest and lowest

temperature? _____

Which two countries have the same highest and lowest temperatures?

_____

• Make up (and answer) five more questions of your own:

a) _____

b) _____

c) _____

d) _____

e) _____

2. This **frequency table** shows the number of people visiting the Science Museum each day for a week.

Monday	Tuesday	Wednesday	Thursday	Friday	Saturday	Sunday
15	35	83	90	42	51	64

a) Use **different** types of graph paper to draw two bar charts to show this information.

For the first graph use one square to represent 5 people.

For the second graph use one small square to represent 10 people.

Don't forget to add the title and to label the axes.

b) Was one bar chart easier to draw? If so, which one?

_____

c) Is one bar chart easier to read?

_____

d) Which is the **most popular** day? _____

e) Which is the **least popular** day? _____

f) Why do you think the Science Museum is more popular on some days?

_____

_____

_____

g) Make up five questions about visitors to the Science Museum that involve calculations – and work out the answers.

_____

_____

_____

_____

_____

Discover

1. Look at the **Venn diagram**.
    Work with a partner.

    - Roll two 8-sided dice.

    - Make the two possible 2-digit numbers.

    - Write the numbers in the correct places on the diagram.

> For example:
>
> I roll 6 and 3
>
> I make 36 and 63
>
> I roll 1 and 4
>
> I make 14 and 41.

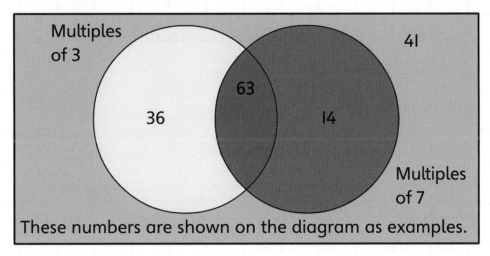

These numbers are shown on the diagram as examples.

- Continue to roll the dice until you start to get repeat numbers.

- How many 2-digit numbers can go in the **intersection**? _____

2. **All about names!**

    Imagine you want to put names of students in your class in this Venn diagram:

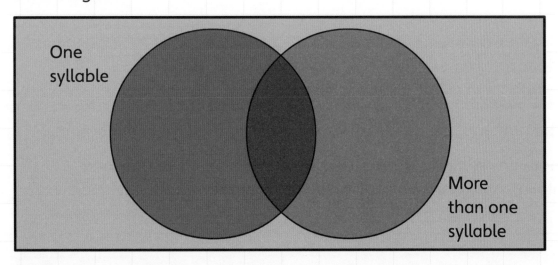

Can there be any names in the intersection? _____

- Explain:_____

A **Carroll diagram** is a better way to **sort** names according to the number of syllables.

- Label this Carroll diagram to answer these questions:

  Does the first name have one syllable or more?

  Does the family name have one syllable or more?

- Add the names of the students in your class to the labelled diagram.

- Are there any students in your class whose names have only one syllable in both their first name and family name?

_____

11C Using Venn diagrams and Carroll diagrams

I. Complete the Carroll diagram for the words in this rhyme:

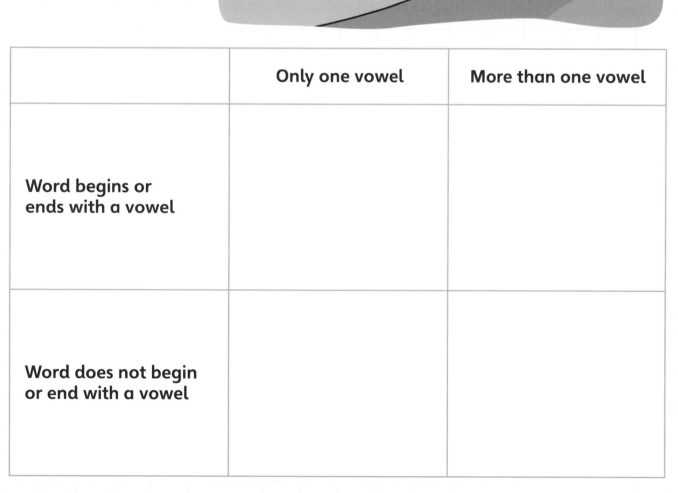

'One, two, three, four, five, once I caught a fish alive!'

'Why did you let it go?'

'Because it bit my finger so.'

'Which finger did it bite?'

'This little finger, on the right.'

	Only one vowel	More than one vowel
Word begins or ends with a vowel		
Word does not begin or end with a vowel		

Remember the vowels are a, e, i, o, u, and sometimes y.

2. Make up a Carroll diagram of your own. Use your own ideas for criteria.

   Try to find at least three numbers to put in each box.

	Criterion I:	Not criterion I:
Criterion 2:		
Not criterion 2:		

3. Here is a shape-sorting Carroll diagram.

   • Think of at least three shapes to write in each box:

	2D Shape	Not a 2D Shape
All straight edges		
Not all straight edges		

11 Handling data

Favourites!

You want to find out the favourite of different categories in your class.

- Choose one category to investigate.

Here are some ideas:

- Favourite lesson at school
- Favourite sport to watch
- Favourite sport to take part in
- Favourite fruit, vegetable or other food
- Favourite TV programme
- An idea of your own ...
- Decide on a sensible question for your **survey**.

_____

_____

- Complete a tally chart and frequency table:

	Tally	Total number

- **Predict** two things that you think you will find out:

_____

_____

- Draw a bar chart or a **pictogram** of your results.

  Take time to make it look as informative and attractive as possible.

Were your predictions correct?

_____

- Write one **surprise** from your data collection:

_____

How could you extend your investigation?

_____

_____

If you have time, carry out this idea too.

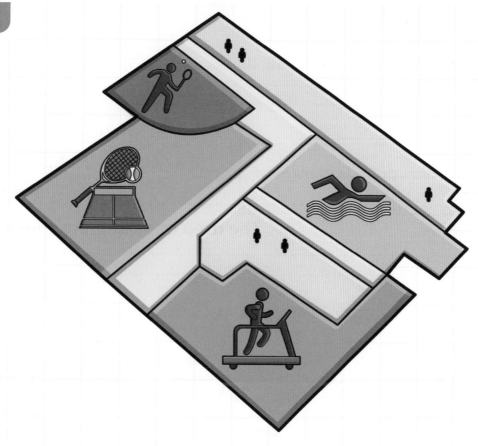

You are the teacher!

- On paper write **instructions** for two new Stage 4 students to carry out this task.

This is the first bar chart they have ever drawn.

- Give them step-by-step instructions.

The Leisure Centre has a:

- swimming pool

- tennis court

- squash court

- gym.

Design a survey to find out how many people visit each of the four areas during one morning.

The new students need to present the findings in a table and as a bar chart.

- Explain how they can do this.

area

_____

_____

_____

The **area** of both these shapes is 2 cm².

base

_____

_____

_____

base of triangle   **base** of cone   square-**based** pyramid

closed

_____

_____

_____

a **closed** shape

concave

_____

_____

_____

a **concave** pentagon

consecutive

_____

_____

_____

14, 15, 16, 17 are **consecutive** numbers.

7, 9, 11, 13 are **consecutive** odd numbers.

−5, −4, −3, −2 are **consecutive** negative numbers.

convex

_____

_____

a **convex** pentagon

coordinates

_____

_____

_____

_____

_____

_____

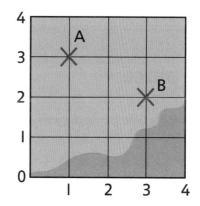

The **coordinates** of A are (1, 3).

The **coordinates** of B are (3, 2).

data

_____

_____

_____

A telephone directory is full of **data** about names, addresses and telephone numbers.

decimal fraction

_____

_____

_____

A **decimal fraction** uses a decimal point:

$$0.5 = \frac{5}{10} = \frac{1}{2} \qquad 0.25 = \frac{25}{100} = \frac{1}{4}$$

decimal number

_____

_____

_____

327

This **decimal** number is made up of:

3 hundreds, 2 tens, 7 units

decimal place

_____

_____

_____

12.56 has 2 **decimal places**.

0.228 has 3 **decimal places**.

3.234677 written to 2 **decimal places** is 3.23.

decimal point

_____

_____

_____

42.6

decimal point

This number is forty-two **point** six.

decrease

_____

Decrease 65 by 15.

Answer: 50

degree

_____

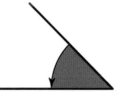

This angle measures **45 degrees**.
We write this as 45°.

eighth

_____

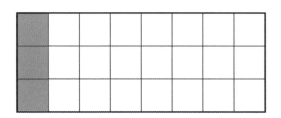

One **eighth** of the shape is coloured.

$\frac{1}{8}$ of 24 is 3.

equilateral triangle

_____

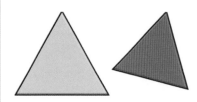

equilateral triangles

factor

_____

3 is a **factor** of 21.

7 is a **factor** of 63.

2 and 7 are both **factors** of 14.

fifth

_____

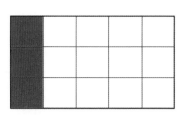

One **fifth** of the shape is colou

$\frac{1}{5}$ of 15 is 3.

greater than (>)

_____

_____

_____

_____

12 > 7
12 is greater than 7.

integer

_____

_____

_____

-4 -3 -2 -1 0 1 2 3 4

negative integers   positive integers

0 is also an **integer**.

heptagon

_____

_____

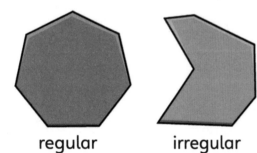

regular heptagon   irregular heptagon

inverse

_____

_____

_____

_____

_____

The **inverse** of + 7 is − 7.

17 + 7 = 24

24 − 7 = 17

The **inverse** of × 4 is ÷ 4.

12 × 4 = 48

48 ÷ 4 = 12

increase

_____

Increase 65 by 15.

Answer: 80

irregular

_____

_____

an **irregular**
hexagon

an **irregular**
triangle

isosceles triangle

_____

_____

_____

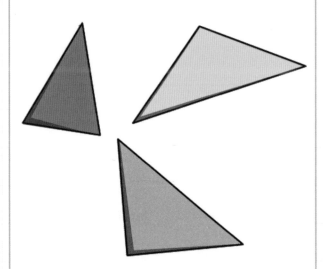

isosceles triangles

leap year

_____

_____

_____

_____

1972   2000   2004   2116

These are all **leap years**.

Usually if you divide the year
exactly by 4 it is a **leap year**.

less than (<)

_____

_____

15 < 20

15 is less than 20.

line symmetry

_____

_____

The kite has
line symmetry.

mass

_____

_____

_____

_____

An astronaut's **mass** is the same on Earth as on the Moon.

But an astronaut weighs less on the Moon than on Earth.

metric unit

_____

_____

Length	Mass	Capacity
millimetre	gram	millilitre
centimetre	kilogram	centilitre
metre	tonne	litre
kilometre		

These are all **metric units**.

millennium

_____

_____

I **millennium** = 10 centuries

millimetre

_____

_____

_____

1000 **millimetres** = 1 metre

10 **mm** = 1 cm

negative number

_____

_____

_____

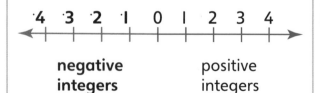

negative integers   positive integers

You can write negative 2 as ⁻2 or −2.

net

_____

_____

_____

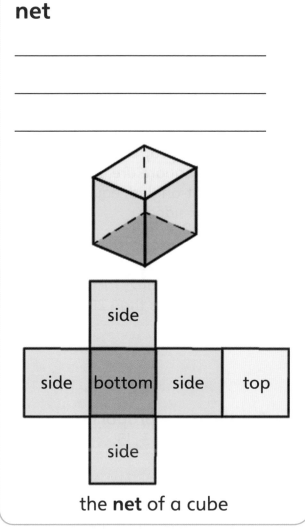

the **net** of a cube

oblong

_____

_____

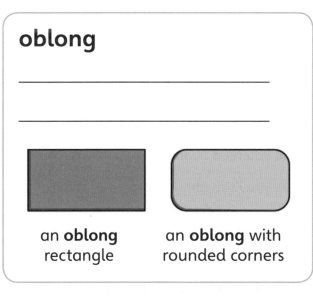

an **oblong**
rectangle

an **oblong** with
rounded corners

open

_____

_____

an **open** shape

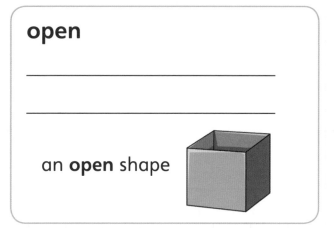

origin

_____

_____

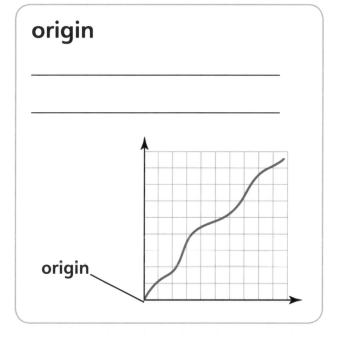

perimeter

_____

_____

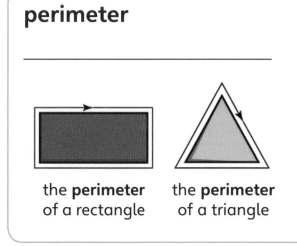

the **perimeter**
of a rectangle

the **perimeter**
of a triangle

positive number

_____

_____

_____

_____

⁻4 ⁻3 ⁻2 ⁻1   0   1   2   3   4

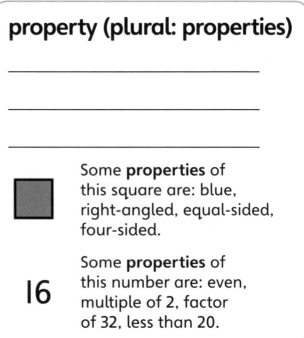

negative numbers   **positive numbers**

You can write positive 2 as
⁺2 or 2.

property (plural: properties)

_____

_____

_____

Some **properties** of this square are: blue, right-angled, equal-sided, four-sided.

16

Some **properties** of this number are: even, multiple of 2, factor of 32, less than 20.

proportion

_____

_____

_____

1:5 is the same **proportion**
as 2:10 or 3:15.

questionnaire

_____

_____

_____

_____

Questionnaire about reading	
Do you read every night?	Yes/No
Do you read fiction?	Yes/No
Do you read about sport?	Yes/No

reflect

_____

_____

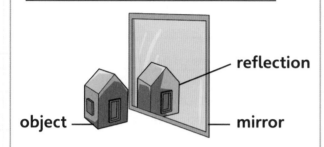

object — | — mirror | reflection

Reflecting is like flipping a shape over.

regular

_____

_____

a **regular** octagon

rotate

_____

_____

_____

The circle is **rotating**.

sixth

_____

_____

_____

_____

_____

_____

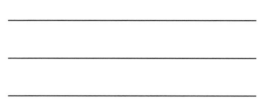

One **sixth** of the shape is coloured.

$\frac{1}{6}$ of 24 is 4.

square centimetre

_____

_____

_____

A 1 cm square has an area of 1 cm²

This shape also has an area of 1 square centimetre.

survey

_____

_____

tetrahedron (plural: tetrahedra)

_____

_____

_____

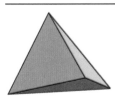

a regular **tetrahedron**

three-dimensional (3D)

_____

_____

_____

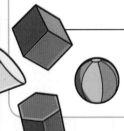

three-dimensional shapes

two-dimensional (2D)

_____

_____

_____

two-dimensional shapes

unit

_____

_____

Litre is a **unit** of capacity.

Kilogram is a **unit** of mass.

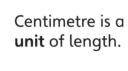

Centimetre is a **unit** of length.